FROM THE CROW'S NEST

FROM THE CROW'S NEST

A Compendium of Speeches and Writings on Maritime and Other Issues

Admiral Arun Prakash, PVSM, AVSM, VrC, VSM (Retd)
Former Chief of Naval Staff, India

National
Maritime
Foundation

Lancer * New Delhi * Olympia Fields IL
www.lancerpublishers.com

LANCER
PUBLISHERS

Published in the United States

by Lancer Publishers, a division of
Lancer InterConsult, Inc.
19900 Governors Drive, Suite 104
Olympia Fields IL 60461

Published in India

by Lancer Publishers & Distributors
2/42 (B) Sarvapriya Vihar
New Delhi-110016

© Admiral Arun Prakash (Retd), 2007

Printed at Sona Printers, New Delhi.
Printed and bound in India.

ISBN: 0-9796174-0-5 978-0-9796174-0-9

www.lancerpublishers.com

CONTENTS

FOREWORD

A very large part of India is a peninsula with a 7,517 km long coastline exposed to the vast expanse of the Indian Ocean. Therefore, India's natural frontiers lie at the far edges of the Indian Ocean, and the countries at its rim are our neighbours. India's rise as an economic power, its role as an important factor in the world's changing power structure and, indeed, its security will increasingly depend on its mastery of the seas around it. Yet, there is so little civic consciousness among our populace about India's maritime potential! As a result our politicians are negligent of the need to harness that potential for India's security and the future well-being and prosperity of our people.

Our poor maritime management of the last four or five decades has left India with a merchant fleet far below the need of its growing sea-borne trade – the share of foreign shipping in carrying India's trade has remained around 70 per cent or more and is likely to increase as India's largely seaborne trade grows – a stifled ship-building industry and decayed and inefficient ports. India is simply without the capacity to exploit the vast and rich resources of its 2.01 million sq. km Exclusive Economic Zone!

Our land frontiers in the north and the west are now secure beyond any serious threat of trespass. The seas around India demand the attention of India's people and their governments: even Chinese thinkers are laying claim to the Indian Ocean as China's next frontier!

India of yore was a thriving maritime power with command of the seas to its east, west and south, which provided mobility and access for its sea-conscious, energetic and outgoing populace to far-away lands. That lent vigour to Indian civilization and brought immense economic and cultural gains to India and its partners in those encounters. Sea-consciousness waned in the Mughal era and India fell victim to sea-faring conquerors from Europe. That story has ended and India and Indians must now look at the Indian Ocean as the area of their opportunity.

Fortunately, we have built up a good Navy, which, though needs to enhance its blue-water capabilities. But maritime power is a combination, besides a capable Navy, of many other things, e.g., a vibrant shipbuilding industry, a large merchant fleet, modern ports with extensive inland infrastructural links, a watchful coast guard and a varied oceanic research establishment, etc. Of the highest importance is well-coordinated maritime management under one single directing and coordinating authority. The absence of such an authority is largely responsible for India's maritime deficiencies in comparison with other maritime powers of the day. A score of Ministries, Departments and other entities dealing with ocean-related activity in the government need to be brought together under one central authority for the fuller and more expeditious harnessing of India's maritime potential.

Good, comprehensive writing on these subjects is in short supply in our country and Admiral Arun Prakash's book – *From The Crow's Nest* – meets a sore need. This highly distinguished officer of the Indian Navy combines a rare sense of history, with a gentle and inspiring vision of India's future as one of the world's great maritime powers. He believes, rightly I think, that India's maritime power is destined to play a key role in India's relations with the Asia Pacific region and, indeed, with the rest of the world. The military in general, and the Navy in particular, have significant roles in any major Powers' conduct of its foreign policy and diplomacy. Chapters in this book reveal, in Arun Prakash's characteristically quiet way, his own significant accomplishments in this area in his time as India's Naval Chief and Chairman of the Chiefs of Staff Committee. The Navy, under his command played a most useful role in the renewal of India's relations with the United States, which awaits recognition.

This multi-dimensional book covers virtually every aspect of India's security, from public-private partnerships for indigenous manufacturing of its military needs, to dilatory and damaging acquisition processes, to military's diverse roles in the nation's life, to the comparative relevance of doctrines and strategies, etc. Our security establishment should pay special attention to what this wise and thoughtful military leader has to say about the necessity of a Chief of Defencte Staff to ensure jointness in our military's planning, decision-making, operations in war and in tendering single-pointed advice to government in matters pertaining to the country's defence.

This is an instructive and easy-reading book on complex matters related to security and international relations. There is much pragmatic wisdom on every page of it for our strategic community and for the general reader interested in India's safety and well-being: I commend it to both.

I hope the book will have a very wide circulation.

New Delhi   Maharajakrishna Rasgotra
1 May 2007

NATIONAL SECURITY

SECURITY AND FOREIGN POLICY
Imperatives of an Emerging India

George Clemenceau's remark about war being "much too serious a matter to be left to the Generals" often comes to mind. I do not take exception to this adage, because today there are people who know just as much about our business as we in the Armed Forces do; and they miss no opportunity to tell us so! What does worry me, however, is the collective sense of indifference with which we have since independence, regarded the study of war, strategy, and national security. This is possibly due to the hubris that is engendered by our ancient heritage and culture, and the consequent feeling that there is nothing much to be gained by a study of others' experiences. But the pity is that we do not even study the fund of politico-military thought available in our own rich heritage of *Vedas*, *Puranas*, and tracts like the *Manu Smriti* and the *Arthashastra*.

The other worrisome facet of our thought process is the pride with which we regard the fact that our civilization has survived the ravages of repeated invasion, conquest and subjugation. Let me quote for you, a verse by the poet Allama Mohammad Iqbal that says:

"Yunan o Misr, Roma sab mit gaye jahan se,

Phir bhi magar hai baki, namo nishan hamara,

Kuchh baat hai ki hasti mitati nahin hamari,

Sadiyon raha hai dushman, daur-e-zaman hamara."

Inspiring words, but one gets the sense that survival by itself is considered a major meritorious achievement. The late George Tanham, who produced the monograph, "India's Strategic Thought" put it succinctly when he said: "The forces of culture and history and

Adapted from an address at the 40th Anniversary Seminar of the Institute of Defence Studies and Analyses, New Delhi, November 11, 2005.

the attitude and policies of independent India have so far, worked against the concept of strategic thinking and planning. As India's environment grows more tense and pressures of every type increase, structures for strategic forethought and planning will have to emerge."

No nation has ever progressed without some sense of national destiny or purpose. In our context however, it seems that it is not Indians but foreigners who have started speaking of India's "manifest destiny". Nothing illustrates this better than the July 28, 2005 edition of the US magazine *Business Week*. The issue is devoted to a minute examination of the Indian and Chinese economies, industry and culture, and paints a cautiously optimistic but balanced picture about the tremendous possibilities that lie in the future for these two countries.

As a self-confident and vibrant India looks towards achieving its manifest destiny in the years ahead, the Indian security establishment will also increasingly need to play a larger role in achieving our national aims. Institutions such as the Institute for Defence Studies and Analyses (IDSA) will, more than ever before, need to present decision-makers with a range of educated policy options on various issues. Indeed the setting up, in February 2005, of the National Maritime Foundation (NMF) which takes its place alongside the Centre for Land Warfare (CLAWS) and the Centre for Air Power Studies (CAPS), means that we now have a complete range of specialist think-tanks, to supplement the efforts of IDSA and other institutions.

As the world awakens to our true potential, India is the flavour of the season and focus of interest internationally with a range of studies and analyses being commissioned to examine virtually every aspect of our country and its potential. The coupling of security and foreign policy perspectives for a resurgent India is therefore not only relevant but, most appropriate in the present environment.

Security goes well beyond strategic and military considerations, to involve political, economic, social, technological and even environmental factors. Emerging concerns also include dwindling energy and water resources, which could become the root of future conflicts. In the post-Cold War era, we have witnessed the tyranny of technologically advanced countries imposing regimes governing technology, space, nuclear energy, and even the environment and human rights. These regimes do not emerge from a consensus or even mutual deliberations, but are discriminatory in nature, and are imposed arbitrarily. They should, therefore, form the underpinning, and provide the context, within which India's defence, foreign, economic and even Science and Technology policies must evolve synergistically.

I would now like to point out a few milestones, which have shaped our security attitudes and structures. Let me begin by casting your mind back over the past fifty-seven years of India's existence as an independent nation, so that we are not "condemned to repeat the past" as George Santayana puts it, by "forgetting it".

It is evident that many of the mindsets, perceptions, and policies that emerged in the early days of our Republic were shaped by the nature of our independence movement. Let me just mention a few of them. I would hasten to add that I am only recalling some facts of history and not making any value judgments.

- First, Mahatma Gandhi's success in achieving independence for India through non-violent means had no parallel in modern history. He was a great man of unique vision and principles, but devotion to the "form" rather than "substance" of the values that he subscribed to led to an idealistic worldview amongst our post-independence leadership. So our policies acquired a *moralpolitik* orientation as opposed to the *realpolitik* of our neighbours. *Moralpolitik* is simply a foreign policy based on morality, which some may consider an oxymoron. This was manifest in Pakistan promptly taking the "low road" to join South-East Asia Treaty Organization (SEATO) and the Baghdad Pact, while we decided to adopt high-minded postures like non-alignment and the quest for universal nuclear disarmament.

- Second, as India's Defence Minister Pranab Mukherjee put it during his June 2005 speech at the Carnegie Endowment, two centuries of colonial rule and exploitation, a large part of it by a group of British merchants who formed the East India Company left Indians suspicious of foreign traders and bred a lingering mistrust of what is now called "globalization". This led to an urge for autarky, or economic self-sufficiency, which we struggled to achieve for over four decades. This was a mixed blessing; for while it did help in building up core industries like steel, space and atomic energy, it also stifled private enterprise in favour of the public sector and led to technological stagnation in several areas.

- If the creation of the Indian National Army in Singapore, and the Indian Legion in Germany out of Indian Prisoners of War (PoW), shook the British Army, the Royal Indian Naval Mutiny possibly sealed the fate of the British Empire. So it is undeniable that the Indian Armed Forces did contribute to accelerating our Independence. The Army also played a major role in persuading a few recalcitrant princely states to join the Indian Union, and helped in handling the massive refugee influx after partition. Notwithstanding this, the Armed Forces were generally perceived as having played only a marginal role in our independence movement, and in some quarters as having been an "instrument of oppression". This led not only to their being marginalized post-independence, and their counsel being largely ignored, but also gave rise to an impression that India could survive with minimal investment in the Armed Forces.

All this is mentioned only to place in perspective our attitudes. Against that backdrop, let us also look at certain post-independence events or junctures, which can be termed as "defining moments" in the evolutionary process of India's national security and foreign policies.

The first of these was October 27, 1947 when barely two months after independence, Indian Army troops landed in Srinagar to stop the invading Pakistani regular and irregular forces from capturing Jammu and Kashmir. Our nascent leadership; both political and military, undertook no overarching analysis of the situation and gave no strategic direction other than to "repulse the invaders!" In a year of fierce but sporadic fighting in inhospitable terrain, the Indian Army gallantly managed to achieve the tactical aims set before it and the Valley of Kashmir remained with us. A haphazard ceasefire line was drawn, and the invaders kept whatever they had captured. The single major outcome of this war was the creation of the so-called "Kashmir problem" which has festered and bled the country for six decades, and currently appears to have become an open-ended issue.

On October 20, 1962 the Chinese army attacked Indian Army positions in North-East Frontier Agency (NEFA) and Ladakh, and the brief but bitter war, which lasted just a month, left us militarily defeated and humiliated as a nation. This debacle while demonstrating naiveté and ineptness in different spheres also exposed our total lack of strategic thinking, planning and vision. A blessing in disguise was the bitter realization that we did not live in utopia and that one could not hope to concentrate on development if you did not ensure a secure environment for the country. And that meant strong, capable armed forces. This marked the end of our post-independence euphoria and recognition of harsh military realities – at least for some time.

Our finest hour, without a doubt, came on December 16, 1971 when 90,000 Pakistani troops surrendered to the victorious Indian forces and which led to the creation of Bangladesh. The nine month prelude to the war saw a diplomatic campaign being mounted in the best traditions of Vedic statecraft, with *sama, dana, bheda* and *danda* all being invoked in turn. The military leadership worked in reasonable harmony, and showed moral courage by buying the time required for a logistics build up. Some imaginative planning resulted in a military campaign, which included every tactic of war; from covert and special operations to air assault, armoured, amphibious, aircraft carrier, submarine and missile boat operations. And yet we faltered in the end-game. No coherent and tangible "war-termination" strategy had been evolved and in the final analysis, it could be said that some of what Zulfikar Ali Bhutto, Pakistan's Prime Minister, lost in the war he recovered in negotiations.

The Peaceful Nuclear Explosion (PNE) in May 1974 marked what should have been another defining moment for India, but in hindsight it can best be seen as yet another, missed opportunity. Already ten years behind China, we should have followed up Pokhran I with a series of tests required to get within striking distance of weaponisation. Perhaps it was a combination of international disapprobation, economic compulsions, and over-caution which made us hold our hand. Whatever the other contributory reasons, lack of a strategic vision certainly lay at the bottom of the tentativeness that marked this "technology demonstration."

The decade of the 1980s saw several major hardware acquisitions by the Armed Forces and it must be flagged as a landmark period. It seemed that India was at last on the way to acquiring the surplus of security assets that would mark it the pre-eminent nation in this region. In the span of about five to six years, the army bolstered its artillery and armour, the navy leased a nuclear submarine, and acquired an aircraft carrier, a squadron each of fighters and maritime aircraft, and the IAF added the Mirage 2000, MiG-29, MiG-27, MiG-23, IL-76 and An-32 to its inventory.

The reality was otherwise. While little change had come about in our strategic thought process, this massive accretion of military might was seen as a serious threat by countries like China, Australia, and others in South-East Asia. We did not bother to rationalize and explain through doctrines or white papers – that was not our style. This build up also coincided with our first forays beyond our shores; into Maldives and Sri Lanka. In the bargain, India began to be suspected of hegemonistic designs and sinister intentions, which just did not exist.

It is now clear that the balance of payment crisis in 1991, which led to the opening up of the Indian economy under the able stewardship of Dr Manmohan Singh, was a defining moment in India's modern history. With the realization that economic growth should underpin India's future relevance in the world, economic policy became the driver for our diplomatic initiatives. Dogma, while not jettisoned totally, took a back seat. A nascent realization began to dawn that our security policy could not exist in splendid isolation and needed to be meshed with our overall foreign policy and economic objectives.

In sum, the 27 years following Independence saw us moving from idealism to a quest for a Western nuclear umbrella, to rejection of the Nuclear Non-Proliferation Treaty (NPT), and then to Pokhran I. The next 24 years could be summed up as a state of "nuclear ambiguity", then "non-weaponised deterrence" and finally in May 1998 came another defining moment: Pokhran II. Whatever our reasons for crossing the nuclear Rubicon and whatever the resulting impact on national security, one thing became clear, at least to students of strategy. It was not the financial liability of maintaining a minimum credible deterrent that was going to be problematic for us. It was the intellectual capital, the time and the capacity required to comprehend arcane nuclear dogma, and to resolutely pursue the minutiae that constitute deterrence, that was going to be difficult to mobilise.

I am, however, happy to conclude my survey of the past on a positive note. Over the recent past, security considerations have started receiving a much higher priority than before, though a great deal still needs to be done in this direction. The structures are now in place, and I would like to think that we are on the way to evolving a sound edifice.

The Prime Minister's visit to the US in July 2005 has the potential to recast our bilateral relations. But we have to see how the reciprocal initiatives on both sides will play out. So this really sums up my list of "defining moments". I will now outline the challenges that lie in store, as India moves forward slowly but surely to take its rightful place in the international order. I intend to highlight just five or six issues of salience.

Firstly, if India aspires to don the mantle of even a regional entity, we have to shed our diffidence, and find not just the ways and means, but the will to project our power overseas. This does not mean that we are going to be aggressors, or to invade someone. We may need to eject intruders from our own island territories, to come to the assistance of our neighbours, to rescue Indian nationals overseas, and as the tsunami showed, to render aid in natural calamities. Or indeed to safeguard our emerging vital interests overseas. Starting with an embryo Rapid Deployment Force, we would need to build adequate sealift and airlift capability to have a credible and sustainable trans-national capability.

Secondly, with oil prices hitting US $70, we have an energy crisis of serious proportions, looming over us. India is currently the world's sixth largest energy consumer, and in 2010, will hit the fourth place. With a great deal of foresight, Oil and Natural Gas Commission (ONGC) Videsh, in addition to a long-term contract for gas supply with Iran, is negotiating with 22 other countries to pursue energy projects involving exploration, development, transportation and refining of hydrocarbons to meet our future needs. If we are going to invest such vast amounts of national resources in locations as far field as Middle East, Africa, Central Asia and South-East Asia it is essential that we take adequate security measures to safeguard our assets and interests against any unforeseen eventualities. This point is closely linked to what I have said above about transnational capabilities, and a rapid deployment force would be ideal for such a purpose.

There is also a move afoot to build up a national strategic oil reserve. To safeguard this oil reserve and to ensure that it is never unduly depleted, our maritime forces would need to be deployed in sufficient strength at strategic locations on the high seas to ensure the safety of our oil traffic in international sea lanes.

Our third security challenge is the rapid and alarming deterioration in the political and economic state of countries, in our immediate neighbours. So much so, that we are in real danger of being completely surrounded by "failed states"; a situation which will be entirely to our own detriment. In addition to their internal turmoil, but possibly related to it, there is a very disturbing phenomenon of growing hostility towards India. This is a challenge that we have to overcome, or it will become not just a millstone around our neck, but an impediment to our rise as a regional power. While political and diplomatic endeavours are going on, we need to establish a link or strengthen existing relations with their militaries. Whether it is confidence building, military aid or even a gentle hint, the Armed Forces can be gainfully utilized for all these purposes.

A fourth area, of vital interest to us, lies in the expanse of the seas — the island nations of the Indian Ocean. Currently, countries like Sri Lanka, Maldives, Mauritius, Seychelles and Comoros are friendly and well disposed to us. However, their security remains fragile, and we cannot afford to have any hostile or inimical power threatening it. These countries are generally in need of military hardware, training or technical assistance and sometimes of help in policing their waters or airspace. Our Armed Forces are always prepared to help, and we are working very closely with the MEA in an endeavour to minimize delays and to meet their needs with promptness. In this context, it might be of interest for this audience to know that the Indian Navy has recently created in Naval Headquarters, a new Directorate of Foreign Cooperation under a two-star Admiral.

The fifth challenge relates to the all-pervasive and omnipresent threat of. terrorism – both within India and globally. This is a phenomenon, which is confined neither by national boundaries, nor by the mediums of land, sea or air. Its sinister tentacles embrace illegal traffic in drugs, arms, human beings, and even in weapons of mass destruction. In combating terrorism lies the biggest security challenge to the Armed Forces, to nations and to the international community as a whole. This challenge is likely to take up much of our energy and resources in the days ahead.

As a responsible nuclear weapon state, our sixth challenge will lie in management of deterrence. Nuclear deterrence, as you all know, lies in the mind of the adversary. To deter someone, you must be able to convince him that the consequences of using a nuclear weapon will be so horrible and devastating, that he should never even contemplate it. Here we are placed in the distinctive situation of being a declared "No First Use State" faced with a nuclear opponent who has in the past threatened first use, and thinks of a seamless "conventional-nuclear war continuum". The only way to make deterrence robust is to ensure that your second strike capability is not only well protected, but that it is also overwhelmingly devastating. Confidence Building Measures (CBMs) certainly have a place in deterrence, as does dialogue and a certain degree of transparency between adversaries.

The final challenge that faces the India's security establishment today relates to transformation of the Armed Forces into a lean, technology intensive, networked and "joint" entity. Jointness and the Revolution in Military Affairs (RMA) are imperatives that the Armed Forces can ignore only at the risk of becoming "dinosaurs" in the current environment. Transformation is a tall order under any circumstances; and the historical experience of other countries indicates that it would need political resolve and direction to initiate as well as sustain such a process.

Conclusion

All in all, there is no doubt that because of India's growing strength, and issues like our prospective permanent membership of the UN Security Council, we will need to integrate our security, foreign and economic policies. A substantial part of the onus will fall upon Track II organisations to provide policy alternatives and options for decision-makers. The Defence Minister, Pranab Mukherjee, while officiating at the foundation-stone laying ceremony of the IDSA's institutional complex had rightly remarked: "While governments have certain responsibilities to discharge, it is the role of the academic and the analyst – what is referred to as the Track II constituency – to provide objective and rigorously analysed inputs on matters pertaining to national security."

Consequently, the role of IDSA and other organisations will grow in importance. However, this will require individuals to have a deep understanding of the subject, which require intense study and rigorous academic analysis. And here let me quote the words of John Adams the Second President of the USA: "I must study politics and war, so that my sons may have the liberty to study mathematics and philosophy."

PLANNING FOR TOMORROW'S NAVY
Challenges in Retrospect

Our racial memory has, with good reason, always been obsessed by the perpetual threat of invasion from the Himalayan passes; the Pakistani marauders who came across Uri in 1947 and the Peoples Liberation Army (PLA) hordes who swarmed down the slopes of Tawang in 1962 only served to reinforce this historical and cultural fixation.

When India gained Independence, those charged with planning for the country's embryonic maritime force were fortuitously, men of vision; and within six months had prepared a ten-year expansion plan for the consideration of the Government of India. The plan was drawn up around the concept of two fleets; one for the Arabian Sea and the other for the Bay of Bengal, each to be built around a light-fleet carrier to be later replaced by larger fleet carriers. This somewhat grandiose plan, which received the approval of both the Governor General Lord Mountbatten and Prime Minister Jawaharlal Nehru, unfortunately failed to materialize. Hostilities with Pakistan in the state of Jammu and Kashmir barely two months after Independence focused the young nation's attention as well as scarce defence resources towards the Himalayas rather than the oceans, and the naval plans were put on the back burner.

Subsequently, during the first few decades post-independence, the Indian Navy (IN) existed in an environment of uncertainty. There was a time when we needed to justify, year after year, our plans, our acquisitions and often our very *raison d' etre* to a skeptical Government. It was only in the 1980s that the Navy's potential as an instrument of state power began to dawn on decision-makers and found a permanent niche in their consciousness.

As the smallest of the three armed forces of a nation beset with a continental mindset, the IN has faced numerous challenges from time to time. This article attempts to provide

Force, Volume 4, Number 4, December 2006, pp.32-36.

a ringside view from the higher reaches of Naval Headquarters (NHQ), *circa* 2003-2006, as well as some personal views and observations relating to force planning issues and processes.

Remoulding Minds

India's emergence as an economic power of global significance and its essential reliance on the sea for energy, trade and projecting influence, is rapidly changing perceptions, and arousing the maritime consciousness of the intelligentsia. While India possesses all the attributes of a potential major power, an inherent cultural diffidence holds her back from assuming the mantle and responsibilities of a regional maritime power. The challenge thus clearly lies in our minds; and re-moulding of perception has been a consistent endeavour of the naval leadership.

Conventional deterrence and war fighting are indeed the bread and butter of navies, but these remain essentially linked to threats, which inevitably tend to wax and wane cyclically with diplomatic activity. Such has been our naiveté and myopia in matters of national security that periodically there emerges a view amongst decision-makers that with 'peace breaking out' all round, the possibility of conflict is diminishing and that defence spending needs to be cut back. On occasions in the past, just as this view was about to prevail, a security crisis has arisen to bring us back from the brink; and so regrettably, we have seen this farcical cycle enacted many times in our brief history.

Of all the armed forces, navies take the longest to build and consolidate; and a growing force like the IN cannot afford to remain hostage to fluctuating security perceptions. The challenge for us, therefore, lies in reducing emphasis on threat perceptions as the sole arbiter in the force planning process, and bringing opinion (within and outside the navy) around to focus firmly a la Palmerston, on India's long-term permanent interests in this context.

A beginning was made in 2004 with the promulgation of the Indian Maritime Doctrine, but the strategic thought process, in order to attain continuity and critical mass, required a degree of institutional reinforcement. The establishment of the National Maritime Foundation (a navy supported non-governmental think tank) in 2005 was the first step in this direction. Subsequently, the creation of the Directorate of Strategy, Concepts and Transformation, the Naval History Division both in NHQ, and the Flag Officer Doctrines & Concepts in Mumbai, have helped to craft an institutional continuum which will hopefully foster doctrinal debate and discussion on maritime issues.

With our maritime interests as the focal point, an exercise was undertaken in 2005 to prepare a Maritime Capabilities Perspective Plan to prioritize the capabilities (as distinct from number of platforms) required to safeguard them in the context of predicted fund availability. It was followed in 2006, by the release of a document entitled, *Freedom of the Seas: India's Maritime Strategy*. This has completed a trilogy of documents, which provides the intellectual underpinning for the Navy's plans and should help to crystallize informed opinion.

The Challenge of Obsolescence

We were fortunate that the seeds of a self-reliant blue water Navy were sown by our farsighted predecessors when they embarked on the brave venture of undertaking warship construction in India four decades ago. Since then, our shipyards have done very well to have delivered more than 85 ships and submarines, many of Indian design, to the IN.

While the hull and even the propulsion machinery of a warship is meant to last for two or three decades, what naval planners dread most is the onset of obsolescence of weapon systems as soon as the ship is launched. This is a very real challenge because a ship may take anything between 6-8 years to construct (in Indian conditions), and since the imported weapons/sensors when nominated for fitment were already in service, they would be 10-15 years (or more) old by the time the ship becomes operational. Thus when the ship completes just half her life, the on-board systems are already over 25 years old and rapidly losing efficacy against contemporary threats.

The latest warship delivered to the navy, INS *Beas*, is stated to be 85 per cent indigenous in content and this is indeed heartening news. But we must face the stark reality that the remaining 15 per cent consists of weapons, sensors and combat management systems, which define the fighting potential of the ship. These systems not only constitute the most expensive component of a warship but, are also most susceptible to obsolescence and have so far remained beyond the capability of the Defence Research and Development Organization (DRDO) as well as the Defence Public Sector Undertakings (DPSU) to design or produce.

It is in a desperate effort to beat obsolescence that the Staff Qualitative Requirements (SQRs) are often pitched at levels considered 'unrealistic', and then not frozen till as late as possible. This has been termed as the classic struggle between what is termed the 'good enough' and the 'best'.

Dependent as we have been, to a very large extent, on various constituents of the former USSR, our shipbuilding endeavours have remained hostage to their opaque, unresponsive and sluggish system of negotiations, contract and supply. This reliance introduces an element of grave uncertainty into the construction schedules and is the single most common cause for cascading time and cost overruns that we have faced in our recent shipbuilding programmes. While the Ministry of Finance may well heap scorn on NHQ and the Ministry of Defence (MoD) for what it considers 'poor programme management', they completely overlook the courageous leap of faith that the Navy has taken by shunning the easier import option and going down the thorny road of indigenous warship design and construction.

Alleviation of this problem has been engaging the attention of the navy for a considerable period, and certain measures have been evolved to reduce its impact. For one, a hard decision had to be taken that the SQRs should be made more realistic, so as to accept current systems, which are 'good enough' to counter extant threats. As a corollary, on the day a unit (ship, submarine or aircraft) enters service, it would be assigned a date for a mid-life update (MLU) a decade or more down the road. This period would permit adequate time for the 'best' contemporary systems to be developed and made available for the MLU.

The ultimate and the only acceptable solution is of course, to become self-reliant and design our own systems, and that constitutes the next challenge.

The Hurdles to Self-Reliance

If there is one lesson that the Indian Armed Forces should have learnt during the past few decades, it is about the hazards and pitfalls of depending on foreign sources for defence hardware (which invariably comes with embedded software). The days of 'friendship prices' are now well behind us, and no matter what the source, we are paying top dollar for everything that we buy in the ruthless international arms bazaar. We must remain acutely conscious of the fact that every time we contract a weapon system or platform of foreign origin, we compromise a little bit of our security because:

- We become dependant on a foreign power for yet one more combat system/platform for its complete life cycle.

- The equipment manufacturer will progressively keep hiking the price of spare parts and overhauls without any rationale or explanation.

- The availability of product support (including spares) will keep declining, till it begins to affect our combat readiness.

- Unless adroitly negotiated in advance, the software source codes will be kept out of our reach to hamper in-house repairs.

Apart from all this we have now repeatedly been witness to the disheartening spectacle of overseas defence purchases being used as political boomerangs and bringing the acquisition process to a grinding halt; thereby affecting the combat capability of the aimed forces.

The obvious panacea for this serious challenge is to encourage our indigenous R&D, as well as industry and to become self-reliant as soon as we can. The Navy's recently established Directorate of Indigenisation has made a good start by focusing on the local production of systems and sub-systems of the Scorpene submarine and the aircraft carrier project and the response from the industry has been most encouraging. But the path of self-reliance is neither easy nor free of pitfalls, as we have learnt from experience.

Over the years, our DPSUs have been manufacturing many systems under so-called 'technology transfer' agreements with foreign firms, but these have resulted only in transfer of 'screwdriver technology' and the assembly of CKD or SKD kits (completely knocked-down and semi-knocked-down), with little or no value addition. That is the reason one has rarely heard of a DPSU producing an improved version of a product after paying huge sums for transfer of technology.

At the other end of the spectrum, the DRDO has often struggled for years at great expense to 'reinvent the wheel' when technology could have been acquired quickly and more economically from other sources. Time overruns and performance shortfalls in many of our indigenous programmes have led to upsets in our force planning process and created operational voids.

In a recent path-breaking initiative the Navy and DRDO have signed a tripartite agreement with Israeli industry for the joint development; and subsequent co-production of a futuristic weapon system for our destroyers of Project-15A. The development cycle of the systems and delivery schedule of the system is planned to coincide so that these frontline ships would be commissioned with a weapon system, which is contemporary, and state-of-the-art worldwide.

An inherent conflict of interest arises from the fact that the DRDO tends to devote much greater resources to technology development and demonstration than to the urgent operational needs of the armed forces. This has often resulted in a mismatch between our critical needs and the priorities of DRDO; driving us towards the import option. There is obviously a need for much better alignment between the aims and objectives of DRDO and the operational missions of the armed forces. In 2004, the Navy had drawn up, mainly for the benefit of DRDO, a 20-year roadmap attempting to forecast the technology requirements that its operational commitments would demand in all three dimensions of maritime warfare. It would be appropriate for the DRDO to take such requirements into account and plan its budget outlay in consultation with the Service HQs.

While the media often lambastes the DRDO (using an equal mix of hyperbole and facts), the Navy has traditionally maintained a symbiotic relationship with this organisation through the three dedicated 'Naval' laboratories to immense mutual benefit. The fact that today the Navy deploys DRDO designed sonars, radars, torpedoes, mines, Electronic Support Measures (ESM), Electronic Counter Measures (ECM) and communication systems, is ample proof of this. We are also funding and supporting the development of the LCA (Navy). However, we have only scratched the surface of the problem and have considerable ground to cover in the arena of self-reliance.

In this context there is a need to clearly understand that India's claim to being a great power or an industrialized nation one day, will ring hollow unless we can acquire the competence to design and build our own ships, submarines, fighters, tanks, missiles and satellites, etc. We also need to accept the likelihood that the first attempt at each of these undertakings may be flawed or even a failure. But had we never attempted to produce a fourth generation fly by-wire fighter, an advanced light helicopter, a main battle tank or an intermediate range ballistic missile (or had we abandoned the projects half-way) it is unlikely that we could have bridged the huge resulting technology gap ever thereafter.

Therefore, a sensible and pragmatic option for the Service HQs today may be to accept the Tejas, Dhruv, Arjun and Agni in their present versions (with certain shortcomings) and dub them 'Mark I'. Then the Services should demand that the DRDO produces 'Mark II' versions of each of these systems and insist that those meet or exceed the SQRs in every respect.

Running the Procurement Gauntlet

Even though it may not appear so, one of the crucial factors impinging on the force planning process is the efficacy of the existing procurement procedures. The absence of a national security doctrine, as well as long-term funding commitment are, by themselves debilitating factors for coherent defence planning in India. But an intractable and ponderous procurement procedure can have a significant impact not just on current, but also future force accretion plans. Witness the stark void in fighter capability, which is currently facing the IAF.

The Defence Procurement Procedure (DPP) has undergone a series of evolutionary changes ever since its inception in 2002, and Pranab Mukherjee released the definitive document in mid-2006. While the DPP has organised and streamlined the processes involved and served to make them as transparent as possible, only time will tell if following these procedures adds to the acquisition time-cycle instead of reducing it. One of the possible pitfalls in this context could be the 30 per cent offset clause, which has been made mandatory for all contracts above Rs 300 crores. Identification of offsets by the bidders and comparison of competing offsets by MoD could both be complex exercises.

A factor, which creates serious impediments in the procurement process, is the current procedure, which subjects each case to the scrutiny of four layers of bureaucracy; the Service HQs, the Department of Defence, the Department of Defence Finance and the Ministry of Finance. After this some cases need Cabinet Committee on Security (CCS) approval. With many queries to be answered, and every file movement taking weeks, if not months, one financial year is simply not enough for most cases to be cleared. Financial authorities have unfettered freedom to examine and re-examine issues, but no corresponding accountability for delays in procurement, which have a cost in terms of national security, lives of troops in the field or monetary loss to the exchequer.

This system defies all logic and it has often been suggested that file movements need to be replaced by a culture of 'collegiate functioning'. Functionaries should process cases by discussion on the phone, walking into each other's offices and calling for regular meetings and discussions. The important observations and decisions can finally be endorsed on file for record.

Choking on Integration

Such a culture of collegiate function can come about only if the wall, which has circumscribed the Service HQs (by terming them first as 'Attached Offices' and now 'Integrated Headquarters' of the MoD), is demolished. The key to efficient functioning lies in bringing the Service HQs inside the MoD and involving them in the national security decision-making process.

Looking at the future, we need to clearly recognise the deep impact that the changing nature of warfare is inevitably having on the force planning process. At the lower end of the spectrum, we need to cater for asymmetric warfare, while simultaneously preparing for conventional conflict. The higher end of the spectrum requires us to plan for credible nuclear deterrence, many elements of which will sooner or later come within the ambit of force planning. It will become increasingly difficult to meet the demands of such planning unless the Service HQs are completely integrated with MoD.

Finally, an essential requirement of long term force planning is the reconciliation of conflicting inter-service demands, prioritization of intra-service plans, and evolution of joint synergies. This may often require adjustments and compromises, which could create controversies. These are best handled within the armed forces by a joint staff with a duly constituted head. The issue of a Chief of Defence Staff (CDS) continues to remain wide open, but whether or not the Government puts a CDS in place, the Services will need to re-engineer themselves to fit into a mould of 'Jointness' otherwise they would have forever abdicated crucial decision-making powers to the MoD.

JOINTNESS IN THE INDIAN ARMED FORCES
Past, Present and Future

Introduction

As it happens, October 6 marks the fifth anniversary of the creation of joint structures in the Indian Armed Forces. Headquarters, Integrated Defence Staff was created on this day. The Group of Ministers (GoM), which oversaw their birth, had recommended that a formal review be undertaken five years after implementation of the new system. It is unlikely that such a review has taken place or will ever take place.

I want to place things in perspective. If there is one lesson that we should take from history, it is that economic and social development just cannot take place in an environment of insecurity. Economic progress has a critical security underpinning that must be clearly recognised and acknowledged. We must also face the harsh truth that weakness can often be a provocation. If we look back at our past conflicts, we will possibly find that by appearing irresolute and indecisive we may have actually encouraged adventurism by our adversaries.

What we have been facing for some years now and will continue to face is best termed as "Asymmetric War"; waged by a ruthless adversary with no-holds barred. The enemy has been planning years ahead and has been cold-bloodedly orchestrating violence in our urban areas amidst the civilian population through a complex and well-organised network of agents and surrogates. This war has many other dimensions too, of which we notice only a few; like demographic invasion, attacks on our economy by pumping in fake currency, inciting communal violence, and undermining the morale and cohesion of the armed forces.

Adapted from the inaugural address at the seminar, "Jointness in the Indian Armed Forces", at the College of Defence Management, Secunderabad, November 24, 2006. The author retired on November 1, 2006.

The most worrisome aspect here is the thought process of the perpetrators of this war; the Pakistan military/Inter-Services Intelligence (ISI) combine. Their calculus runs seamlessly from sub-conventional or unconventional to conventional warfare, and then on to nuclear conflict; the whole paradigm working in tandem with clever diplomatic posturing.

We, on the other hand, have kept the different aspects of this conflict strictly compartmentalised and hence our response to the asymmetric war is fragmented and often disorganized. The main reason for this is; that our national security establishment deprives itself of the benefits of holistic thinking and action, which would result from regular consultation with the armed forces.

The reasons for this incoherent and flawed approach are two-fold. Firstly, the three Services being insufficiently integrated, function as separate entities, and therefore lack clout in important security forums. And secondly; that the Service HQs remain organizationally external to the MoD and are consequently de-linked from all-important decision-making processes, and often from consultation on important security matters.

Integration and Jointness are matters of great importance that merit the time and attention of the security establishment as a whole, and specifically of the armed forces. There are just two limited objectives of this essay. Firstly to provide a wider perspective on integration and jointness and to show that we have neither studied nor learnt from history. And, secondly, to provoke some soul-searching amongst ourselves if possible.

I start at the beginning, with the genesis of jointness in its country of birth - the USA. And after that I will trace the evolution of our own endeavours in this field.

A BRIEF HISTORY OF JOINTNESS

The Advent of Air Power

World War I was the last conflict in which the services, everywhere, could maintain their autonomy without sacrificing combat efficiency. Air power changed all that forever. Shortly after the armistice in 1918, both the US Navy and the US Army saw the writing on the wall and developed significant air components. However, the doctrinal outlook of the two services could not have been more divergent.

Naval tacticians saw aviation as just one more component of an integrated approach to maritime warfare, involving ships, submarines, and now aircraft. In contrast, non-naval tacticians, led by the US Army General Billy Mitchell, argued that air power should be used, not as tactical support for ground or maritime forces, but rather as a means for defeating the enemy by destroying his population and centres of production.

Polarization of Doctrines

As World War II loomed large, each service propounded with fervour, a specific theory of warfare, which best suited its interests. The US Army viewed successful ground operations as the pre-requisite for victory, the Navy saw control of the seas as critical to global dominance, and the Army Air Corps was convinced that only massive aerial bombardment could pound the enemy into submission. The divergent operational perspectives also gave rise to a polarization of attitudes as far as jointmanship or centralization of control was concerned and this is quite educative in the Indian context, because it shows that armed forces everywhere, develop an outlook, which is shaped more by their narrow, parochial interests, rather than larger national security considerations.

Beginnings of Jointness – The National Security Act, 1947

The US administration, for reasons, which are very akin to those which influenced our own system, kept the services totally apart and independent of each other well into World War II. But by the end of the war it was obvious to the political establishment that growing dissension amongst the Armed Forces had led to many instances of confusion and discord in operations, and urgent reform was called for.

At this juncture, three other factors emerged which precipitated matters. Firstly, the Army Air Corps broke away to become a separate service – the US Air Force. Secondly, it became clear that atomic weapons would be the arbiters of future wars, and that aircraft would be the preferred choice as a vehicle for their delivery for the foreseeable future. And lastly, everyone realized that the lion's share of defence dollars would go to the service, which wielded the A-bomb.

The US Army proposed unification of the services, and President Truman initiated a wide-ranging debate amongst Senators and Congressmen. As per US custom, service officers were permitted to testify and place their views before Congress. Active use was also made of the media by the services, to push their points of view. A great deal of political activity and debate resulted in the US National Security Act of 1947.

This was a seminal piece of legislation, and established the US National Security structure as we see it today. The Act created the National Security Council and the Central Intelligence Agency (CIA), as well as an apex body called the "National Military Establishment" (NME) headed by a political appointee to be known as Secretary of Defence. It also accorded formal recognition to the institution of Joint Chiefs of Staff (JCS) as a body of equals. However, no sooner had this legislation been passed that fierce debates broke out between the Navy and the USAF about "Roles and Missions", and between the Navy and Army about the relevance of the Marine Corps.

Basically all the rhetoric boiled down to getting a larger slice of the budgetary cake, and this is where the flaws of the new system emerged. The powers of the Secretary of Defence had not been clearly defined, and he depended on the JCS for advice, especially regarding budget priorities. The JCS, on the other hand, was a body of equals and each Chief preferred to represent the interests of his constituency, rather than looking at larger national security interests.

Therefore, just like our own COSC, no decisions of importance ever emerged from the JCS. If pressed hard, they would produce bottom line, consensus decisions on matters of little consequence. This led the Congress to press for another set of reforms in 1956, which represented a dramatic shift towards unification and a setback for the US Navy.

Goldwater-Nichols: The DOD Reorganization Act of 1986

Between 1958 and 1980, there was relative stability in the US JCS structure, but it received adverse attention because of frequent problems, failures, and fiascos in US defence policy-making and in military operations. Apart from the Vietnam experience, there were other crisis situations whose inept handling led to strident criticism of the JCS system. Among these were the seizure of the intelligence gathering ship USS *Pueblo* by the North Koreans, the failed Iran hostage rescue, the bombing of the Marine barracks in Beirut and the chaotic Grenada invasion. Over the next four years, an intense public debate raged on the Hill, culminating in the passage, by the Congress of the Goldwater-Nichols Department of Defence Reorganization Act 1986.

This Act, described as one of the landmark laws in American history, encompassed the following changes in the US National Security structure:

- It elevated the Chairman JCS to be the Principal Military Adviser to the President, but it permitted a dissenting Service Chief to represent his views to the civilian leadership.

- It maintained the Chairman's status as the highest-ranking military officer, but precluded him from exercising command over the Joint Chiefs or any of the Armed Forces. He was charged with assisting the Secretary of Defence to prioritize the budget, and preparing strategic plans.

- It required that all military forces be assigned to the Cs-in-C unless required for training. Units assigned to a C-in-C could not be re-assigned without permission of the Secretary of Defence. The chain of command was delineated as running from President to Secretary of Defence and thence to the Commanders-in-Chief.

- It implemented measures to ensure that high calibre officers were assigned to joint posts, and created a "joint specialty" which required courses of study at specified military institutions.

- In pursuance of the above objective, it mandated that certain stipulated service in joint organisations was mandatory for promotion to Flag rank in all three services.

The Goldwater-Nichols Act represented the most sweeping changes in the US National Security arena since 1947, and is seen to have had a most beneficial impact, as seen from the performance of US Forces in Panama, Somalia, Kosovo, Iraq and lately in Afghanistan.

Goldwater-Nichols in the Indian Context

It was necessary to dwell at length on this subject in order to illustrate that virtually everything that took place in the USA has happened or will happen here, including the debates on Jointness, requirement of a CDS, air power and control of nuclear forces. It is, therefore, necessary to study their experiences, and find lessons relevant to India. We also need to clearly understand a few other factors.

In the current environment, our polity has its hands full with issues of regional, social and electoral importance. National security per se does not figure prominently on the agenda of the political establishment and it is difficult for them to get fully involved in defence issues of a complex nature and find a resolution. At the same time, the Armed Forces (like their counterparts worldwide) being conservative and essentially "status-quoist" will not be able to bring about any significant change on their own.

We therefore need to be realistic, and face up to the fact that at this moment, jointmanship in the Indian context is just skin-deep and to an extent, cosmetic. It should also be clear to all concerned that the services will compete with each other fiercely for what they perceive as their core interests; be it creation of new formations, increase in higher ranks, or their share of the budgetary cake. This is how we have grown up and this is what the services expect of their Chiefs.

So the question arises; if the services themselves are unable to synergize and unite, and the polity is unwilling to intervene, are we doomed to forego the economy, efficiency and operational synergy that emerge from jointmanship forever? This is a rhetorical question, and I address it as much to myself as to the audience. Before addressing it, let us cast our eyes inwards and see how we in India went about a somewhat similar and long overdue exercise of reform in the higher management of defence.

A REVIEW OF DEFENCE MANAGEMENT IN INDIA

Historical Context

In order to obtain a correct perspective, I will delve into a bit of history once again. In very general terms, the British Commander-in-Chief (C-in-C) in India looked after operational matters and rendered advice to the administration on military issues, in his capacity as a member of the Governor-General's Council. The Governor-General had his own military department, headed by a military member of the council, to convey his directions to the C-in-C.

This system, with some variations, served the purpose till the early years of the 20th century, when the armed forces in India began to be seen as an extension of the British war machine and a useful instrument of imperial strategy, and their development and higher direction guided by British rather than Indian interests.

Post-Independence Formulation

In 1947, India was fortunate to have, at the helm of affairs, two very experienced military leaders, namely Lord Mountbatten as the Governor-General, and Lord Ismay as his Chief of Staff. The Government of India asked them to apply their minds to the evolution of a system of defence management for the newly independent country. Highly disturbed conditions prevailed in India at that juncture, and the Armed Forces, like the country, were about to be partitioned. Under these circumstances, these two gentlemen did not consider a radical reorganisation of the armed forces appropriate.

Lord Ismay, who had earlier been invited by the USA to advise on national defence, therefore, pragmatically recommended a system which encompassed a Chiefs of Staff Committee as well as a series of other committees which would ensure supremacy of the civil over the military, enable coordination between the services, and provide for quick decision-making with minimum red tape. These committees had civil servants as members and their decisions carried weight with the Ministry of Defence (MoD).

This framework was accepted by a national leadership quite unfamiliar with the intricacies of national security management, and preoccupied with the manifold pressing issues of Independence and partition.

The Bureaucracy Strikes

Ismay's new system was a workable proposition and could have evolved and integrated itself with the government with passage of time. However, personalities and circumstances at that juncture intervened to distort the concept of 'civil supremacy' and to give it the interpretation of 'bureaucratic control'. This was done by the simple expedient of designating the Service Headquarters as "Attached Offices" of the MoD instead of giving them the status of independent departments of the Government that they rightfully should have become.

The young military leadership of that period was either too inexperienced, or too preoccupied to determinedly oppose this fait accompli that they were presented with.

Over the years, by design or by default, the status of these committees, the powers of the Service Chiefs, and the administrative effectiveness of the Service HQs have all badly eroded, and the services now find that they are adjuncts to, but completely outside the Ministry of Defence.

In their own sporadic and disjointed manner, the services did try to fight this iniquitous dispensation, from time to time. However, it took half a century of frustration, a humiliating defeat by the PLA, numerous other close calls, and finally the near disaster of Kargil to trigger some change.

Post-Kargil Developments

The Kargil Review Committee formed under the chairmanship of former civil servant and defence analyst, K. Subrahmanyam, went into the events leading up to, and the conduct of the Kargil campaign. The findings of the committee were a scathing indictment of the deficiencies in our intelligence services, the higher defence organisation and the management of our borders. The KRC urged a thorough and expeditious overhaul of the national security system, insisting that the bureaucracy should not undertake it.

The Committee led to the formation of a group of ministers; which in turn commissioned four task forces to undertake a critical examination of various aspects of national security. The task force on defence was the one headed by the former Minister of State for Defence, Arun Singh and of which I was a member.

This body was charged with; essentially, a critical examination of existing structures for management of defence, against the background of the RMA and our new status as a *de facto* nuclear weapon state, and asked to suggest changes for improving management of defence. It was also tasked to look at the interface between the MoD and Service HQs, as well as the need for integration between the services.

Labours of the Task Force

During the four months of deliberations that were undertaken by the Arun Singh Task Force, it was an uncanny experience to see history repeating itself. The positions taken and arguments put forth were almost identical to those that have just been described in the American context.

As in the USA, a very basic issue, which kept raising its head was that unless the roles and missions of each service were clearly delineated, it would be infructuous to undertake any study related to changes in defence management.

Here, the role of aviation and control of aviation assets were the underlying bones of contention. There were extended discussions on the need for and extent of integration between the Services and MoD, and within the services. And of course, the most serious item of contention turned out to be that relating to the constitution of a CDS.

The main plank of those favouring status quo was that the Chiefs of Staff Committee (COSC) system had worked quite well for the past fifty years, and therefore we needed neither a CDS, nor integration with MoD nor any further inter-service integration.

The creation of the Andaman and Nicobar Command (ANC) did not raise much debate, as it was virtually a gift from the Navy, and a net gain for the other two Services. In the context of Strategic Forces, there were serious differences whether they should be under single Service control or form a joint command. If you cast your minds back to what I had stated earlier about the US scenario, you will note with satisfaction that there are some things in life, which never change!

Task Force Recommendations

The task force made a large number of recommendations, of which the main ones relating to higher defence management were:

- That the Service HQs be re-designated as "Associated Offices" in the MoD, and substantive delegation of financial and administrative powers be made to them.

- That the existing COSC be enlarged by the addition of a CDS who will be the permanent chairman, and a VCDS who will be the Member Secretary. An alternate recommendation was also offered.

- The CDS was to be the Principal Military Adviser to the Government of India. He would not exercise command over any of the Chiefs or Forces other than those placed specifically under his command.

- Two joint formations; the Strategic Forces Command (SFC) and the ANC were to be established.

- The ANC was to be commanded initially by a three-star naval officer, and then the post be made tenable by officers from all three services. CINCAN to have under command, all Army, Navy, IAF and CG assets in the islands. He would report directly to the CDS/Chairman COSC.

Jointness Today

The higher defence organisation that we inherited as a temporary expedient to tide over the turbulence of Independence and partition stayed with us for 54 years. The Kargil Review Committee did lead to some changes, which were partially implemented in

2001. However, the current organisation, except for three joint institutions and some devolution of powers, essentially remains 'Ismay Mark II'.

It must be admitted that even today, externally, there is strong opposition to further autonomy and devolution of powers to the armed forces HQs, and any integration of the services into the main body of the MoD. Within the services there appears to be a dichotomy. In the field, middle and senior ranking officers appear enthusiastic about jointness, but in Delhi one hears only deafening silence and meets subtle but stubborn resistance to taking jointmanship beyond mere symbolic gestures.

And of course the issue of CDS, for different reasons, finds fierce opponents within and without the services. But we need to ask ourselves if we have taken the right positions over a vital issue?

Today, in the throes of globalization, India's economic resurgence as well as her emergence as a regional power have their own far reaching implications. We do not believe in 'spheres of influence' or tenets like the 'Monroe Doctrine', but in terms of economy, working population and market dynamics India is fast approaching a certain 'critical mass'. All these factors have their own compulsions and we will find that they create vital national interests, which we must safeguard.

Overseas trade, energy security, a large merchant shipping fleet, island territories, the far-flung Indian diaspora, and real estate are being acquired by ONGC worldwide for oil and gas exploration. These are some of the vital national interests that require us to shed our traditional insularity and look outwards. The Africa summit held in Beijing in November 2006 should be an object lesson for us on how a nation's foreign policy should merge seamlessly with economic and strategic considerations to safeguard her long-term national interests.

Since the 1980s we have demonstrated the unique ability of rendering assistance to our friends without strings or ulterior motives. After the tsunami relief operation and the Lebanon refugee crisis, our neighbours will now have renewed expectations from us. In this context, the Navy's thought process and operations have traditionally been transnational, and now the IAF with its extended reach also seems to be speaking the same language. The Indian Army too, with its vast experience of UN operations and bilateral exercises should have no difficulty in undertaking out of area operations.

However, we unfortunately remain in a state of denial, and one of the main arguments in favour of status quo in our higher defence organisation is our so-called pre-occupation with internal security issues. It is said that the concept of integrated operations, theatre commands and a CDS is applicable only to those nations who contemplate expeditionary operations. And that seems to exclude us.

One of these days we will have to ask ourselves some hard questions. How long are we going to remain bogged down and inward looking? How long are we going to retain our cultural diffidence and reluctance to assume the responsibilities of a regional power? The day we can answer these questions, we will also need to confront the issue of Jointness and our higher defence organisation.

Conclusion

In conclusion, let me quote three of the main impediments to progress of jointness in our armed forces.

- Firstly, there is what I call the 'ostrich syndrome', which says that we have always been joint because of the National Defence Academy (NDA), Defence Services Staff College (DSSC) and the National Defence College (NDC), and nothing more can or need be done in this area.

- Secondly, jointness will inevitably bring with it, loss of 'turf' and authority, and there are narrow single service interests, which militate against this prospect.

- Finally, change brings turbulence and status quo is always comfortable. So many voices say: "if it ain't broke why fix it?"

While acknowledging the separate identity of each service, and respecting the divergence of views, we must look around at what is happening in the rest of the world, and remain extremely wary that for short-term parochial gains, we do not cause harm to the long-term interests of the armed forces and our nation.

Challenges of Defence Planning

Defence planning even in the most conducive of environments presents many problems and challenges, but in a developing country, which also happens to be a boisterous and often disorganized democracy, the challenges are multiplied manifold.

The debate about "guns versus butter" is probably as old as mankind, and I am sure that the same old arguments have been used on either side of the divide, over the centuries. The response of nations to such conundrums arises from many background factors. In the case of India, attitudes have possibly been conditioned by not just its geographical location but also its historical experience and socio-cultural tradition.

In this context, let me quote an outsider's view of our strategic thought process. An American analyst has stated, and I quote: "Through much of its history India has remained on the strategic defensive. Over the centuries, armies from Central Asia and Persia invaded the subcontinent by land, and Europeans have invaded by sea. Rarely did the Indians succeed in repelling invaders, and they seemed not to have learnt from their experience. The small Indian states concentrated more on fighting each other or assisting the invader against a rival than turning back the invader collectively." The analyst concludes by saying: "The forces of culture and history, and the attitudes and policies of independent India have, so far, worked against the concept of strategic thinking and planning, but as the environment grows more tense, and pressures of all types increase, structures for strategic planning will have to emerge."

This has been proved true not just by our military debacle in 1962, but on many occasions, thereafter. As a nation, we tend to lapse periodically into pacifist day-dreams wherein we imagine that just because we have no expansionist designs or ill-will towards others, the same benign environment prevails universally.

Adapted from the Keynote address delivered at the seminar "Challenges of Defence Planning" jointly organized by the Department of Defence Finance and IDSA, New Delhi, July 15, 2006.

The corollary that emerges from this utopian vision is that we should concentrate our energies and resources exclusively on the developmental process, and adopt a minimalist approach towards national security Unfortunately the reality is quite the contrary; because others are more pragmatic and will always act on the basis of their self-interest.

There can be no doubt that our priorities should be firmly focused on uplifting our sick, hungry, and illiterate masses from their crushing poverty. But what is frequently overlooked is the fact that security and development have to go hand in hand, and this is a lesson that emerges repeatedly from India's own recent history. Unless we can create a secure environment for ourselves, which will insulate us from external intervention, the vital process of human development will not be allowed to proceed. National security has to be therefore treated like an insurance policy in which adequate investment must be made for securing peace today, so that future generations can experience prosperity tomorrow

Economists have also helped to muddy the waters by often proclaiming that national security and economic development are mutually exclusive. Empirical evidence, however, shows that defence expenditure, especially in democracies, has a beneficial impact on growth and investment as long as it is kept within a prudent limit of up to about 5 per cent of GDP. As a matter of historical interest, defence allocations in India were as low as 1.8 per cent of GDP for the first 12 years after independence, which has often been cited as a reason for our debacle in the Sino-Indian war of 1962. For the next 25 years till 1987, money for defence was forthcoming at an average annual rate of 3.05 per cent of GDP, and there is enough evidence to show that this had a beneficial impact on industrial and economic growth rates.

My own view on the subject of "guns versus butter", is that weakness can paradoxically become "provocative" because it leads to a breakdown of deterrence; and there is no better example of this than the India-Pakistan equation. It can stated quite candidly, that we may have ourselves tempted Pakistan into repeated adventurism by our demonstration of inadequate resolve, lack of determination, and inability to demonstrate a national will.

We do not seem to acknowledge this, but even while we think we are at peace, an "asymmetric war" is being waged against India. There is obviously in place, a long-term strategy, which provides men, material, finance and overall direction to the campaign of subversion, sabotage and bombings which takes place before our eyes, so frequently; and Akshardham, Delhi, Ayodhya, Varanasi, Srinagar, and Mumbai are perhaps just the beginning.

This brings me to the first of the challenges that face military staffs today: what should be the genesis or starting point of defence planning? The first essential pre-requisite, obviously, is the formulation and articulation of a comprehensive National Security Strategy, which would then define salient National Objectives. From these would emerge the broad contours of a supporting military strategy, which would not only bring focus to, but perhaps also circumscribe the defence planning process.

In fact, we are only now reaching a stage where National Security has begun to be treated as a vital seamless issue cutting across many disciplines, ministries and departments. Realization is also dawning that it must be studied in a cohesive manner to throw up policies and strategies. But it is not yet our style to articulate any security related doctrines, which would clear up ambiguities and bring our priorities into sharp focus in the planning process.

The defence planner does, therefore, start with a handicap and tends to grope a bit for direction. However, this does not mean that we plan in a vacuum; we have the Defence Minister's Directive as a start point. From here, scenario-building exercises are undertaken and assessments made so that the planning process proceeds unhindered. But what is our instrumentality for planning?

In 1947, the Government of newly independent India, put in place a system of higher defence management (HDM) based on the advice of Lord Ismay, Mountbatten's Chief of Staff. The system encompassed a COSC and a set of functional committees, which would ensure supremacy of the civil, over the military, enable coordination between the Services and provide for quick decision-making with minimum red tape. The system has survived for six decades, but suffered great loss of efficiency over this period. This is essentially due to the fact that the Service HQs instead of being integrated into MoD were put outside the pale, and as the gap grew larger, they have lost their influence and effectiveness.

In 2001, based on the recommendations of a Group of Ministers (GoM), certain changes were brought about in the HDM in an endeavour to enhance efficiency and attain integration; amongst the Services as well as between Services and the MoD. Although a far reaching transformation had been envisaged, the organization that finally emerged was something of a halfway house and it is difficult to say whether we are better or worse off than before. It is, however, obvious that there is a continued lack of clarity or consensus within our polity on the issues of integration of the Armed Forces and the institution of a CDS.

These issues may not affect the day-to-day functioning of the Armed Forces, but one area in which this uncertainty does have an adverse impact is force planning. A key requirement of long-term integrated planning is the reconciliation of conflicting inter-Service demands, prioritization of plans, and evolution of joint synergies. This may often require "robbing Peter to pay Paul" which creates controversies; and the best entity for handling them is certainly not a parttime Chairman Chiefs of Staff Committee, who has no powers to overrule his colleagues. So we often find that an exercise best done in-house by the Services has to be eventually undertaken by the MoD or MoD (Finance).

Another related issue is, that the Chiefs currently bear on their shoulders, the final responsibility for both "command" and "staff" functions of their Service. The "staff" function requires them to plan force levels, and equip as well as train their forces to fight future wars. But not being part of the MoD, they can only make recommendations, and are often unable to wield the influence necessary to maintain current forces, or to shape them for future conflicts. This issue of separation of the "command" and "staff" functions of the Service Chiefs will crop up in a big way as and when we contemplate creation of Theatre Commands under a CDS.

The lack of a long-term funding commitment has always been an impediment for coherent defence planning. Perhaps the absence of a national security doctrine is one of the underlying reasons why such commitments have never been forthcoming for our plans. The Service HQs have frequently, in the past, taken great pains to produce force level and infrastructure perspective plans. These have been scrutinized by the MoD but since no formal approvals have ever been accorded, they were relegated to the status of academic studies.

In our new scheme of things the genesis of the acquisition process lies in the Long Term Integrated Perspective Plan (LTIPP). Once approved, the LTIPP should lead to Five Year plans, which in turn generate annual acquisition plans. We have so far, never received an in-principle approval of the 15-year perspective plans, nor have our 5-year plans (except for the 9th Plan) ever been underwritten by the Ministry of Finance. This has reduced the defence planning exercise to a hurried and somewhat ad hoc annual ritual. With a long drawn out selection process, and even longer gestation periods, equipment acquisition has become problematic in the absence of assured long-term funding.

However, there appears to be good news on the horizon. The LTIPP for 2002-2017 has been under examination of the MoD. There are indications that the MoD will accord approval to the Twelfth Five Year Plan. While this is a major step forward, it will become meaningful only when the Ministry of Finance sets its seal of approval on the plan.

What about the acquisition process itself, because it has a significant upstream impact on current as well as future plans? The new Defence Procurement Procedure (DPP) has undergone a series of evolutionary changes right from its inception in 2002, and should soon emerge in its definitive form. While the DPP has organised and streamlined the processes involved, and served to make them as transparent as possible, it is likely that following these procedures will add to the time cycle involved in acquisition. A good example is the 30 per cent offset clause mandatory for all contracts above Rs 300 crores. There will be many instances where identifying suitable offsets would itself become a complex exercise.

Let us take another major imponderable which presents an area of difficulty in our defence planning process; our dependence on foreign sources for weapons, systems, and equipment. I will give you an example. In the Navy, we are very proud of our home grown shipbuilding programme; because we have on order today, a large number of ships ranging from patrol boats to landing ships, destroyers, frigates, and submarines to an aircraft carrier. But while we build the hulls, engines and a large number of other systems in India, the weapons and sensors are still imported. Invariably, there have been interminable delays in negotiations, contracting and supply of these items, and an accompanying rise in prices. These delays have had a cascading effect on the shipbuilding process with a consequent rise in construction costs.

That is not all, because apart from uncertainties in delivery and cost, sometimes even the reliability of the imported systems is low, and coupled with poor product support; it leads to further turbulence in the planning matrix. The answer is obviously to encourage our indigenous R&D as well as industry and become self-reliant as soon as we can. But there are many pitfalls down this road too.

Firstly, most of the "technology transfer" deals that we have so optimistically concluded over the years have often resulted in transfer of nothing but "screwdriver technology" and assembly of SKD or CKD kits. That is why we have rarely heard a PSU talking of producing a Mark II product after paying through our noses for so-called "ToT".

At the other end of the spectrum, we have often attempted to "reinvent the wheel" when technology could have been acquired quickly and more economically from other sources. Time overruns and performance shortfalls in many of our indigenous weapon programmes have led to upsets in our force planning process, and often created operational voids.

An inherent conflict of interest exists in this area because R&D (perhaps with some justification) tends to devote much greater resources to technology development than to the operational needs of the armed forces. This has often resulted in a mismatch between our critical needs and the priorities of R&D, driving the Armed Forces towards the import option. There is obviously a need for better alignment between the aims and objectives of our R&D and the operational missions of the Armed Forces.

We have now taken two decisions in this context. Firstly, we will accept the weapons or systems created by R&D, with their current performance limitations, and designate them as "Mark I". The Service will then generate a Staff Requirement for a Mark II version of the equipment and expect R&D to meet it in full, within a specified time-frame.

As far as acquisition of technology is concerned, the relevant advanced technologies should be identified by the Service HQ preferably at the nascent stage and negotiations undertaken with the foreign source for joint development as well as collaborative production. The Navy, DRDO and a foreign company have recently signed a tripartite contract on these lines for the joint development and production of an advanced system wherein DRDO scientists, and Navy engineers will work alongside the foreign experts, both in India and abroad.

Our planners have also been revisiting the time-honoured "bean count" methodology of force planning by numbers. Increasingly, they are looking at the "capability" required that is to be delivered at the scene of action as the start point of the process. By examining this requirement against the backdrop of the enormously enhanced reach, firepower and effectiveness of modern systems as well as their huge costs, they derive affordable numbers, which will meet the task. The Indian Navy has, on this basis, recently evolved a Plan by which it is hoped to deliver a "right sized", affordable and robust maritime force by 2022.

Looking ahead, we must clearly recognise the deep impact that the changing nature of warfare will inevitably have on the defence planning process. At the lower end of the spectrum, we will need to cater for asymmetric warfare waged by non-state actors, while simultaneously preparing for conventional conflict, which will, now and forever, be conducted under the shadow of a nuclear threat. The higher end of the spectrum will require us to plan for nuclear deterrence against those who may threaten us in that arena. Sooner or later, many elements of our nuclear deterrent will also come within the ambit of the force planning process in Service Headquarters, and these would require comprehension and skills of a different order.

In all our planning, we can never afford to lose sight of the tremendous efficiency, and economy that can flow out of the synergy delivered by Jointmanship. Whether or not the Government puts a CDS system in place, the Services will need to re-engineer or even reinvent themselves to fit into a "Joint" mould. To this end, all of us need to think seriously about initiating a process of "transformation" which will bring about the profound changes in training, maintenance, operations and logistics necessary to meet the compulsions of jointmanship, the demands of the RMA and the need for networking.

India is poised on a growth trajectory that promises to propel it into its rightful place alongside the major economic entities in the world. The realization is slowly but progressively dawning on our intelligentsia that economic growth may turn out to be meaningless unless underpinned by concomitant military power, which will guarantee national security. Prime Minister Manmohan Singh mentioned to the Combined Commanders in October 2006 that if the growth rate was sustained at 8.5 per cent, a defence expenditure allocation to 3 per cent of GDP could be visualized.

Already, we are dealing with a defence budget, which would have seemed of "astronomical" proportions just a few years ago. Should the PM's prediction come true, enormous amount of funds would be pumped into the defence establishment and we would require a very efficient, effective and perhaps different system of planning and acquisition. The time to start thinking about all this is now.

OUTSOURCING OF DEFENCE PRODUCTION
Opportunities and challenges

The Kelkar Committee report, as you know had focused on "strengthening self-reliance in defence preparedness", and this has logically led them to recommend measures which will create synergies between the private and public sectors, thereby involving the country's best industrial resources in defence capability building.

There used to be a school of thought in the Armed Forces, which said: "my job is to defeat the nation's enemies, and I don't care if I do it with an Indian bullet or an imported one. Just make sure that I have a bullet that works, and that its there when I need it." Now this may sound incredibly shortsighted, but it reflects the soldier's impatience with delays, which had become endemic in our procurement system. But over the years, we have become wiser and learnt through bitter experience, that in the long term, it is only the metaphorical "Indian bullet" which will actually defeat the enemy and truly save the nation. The scrapping of the Industrial Policy Resolution of 1956 and opening of the defence sector to private industry is very much a part of this process, which will give us the "Indian bullet".

There are many who criticize India's early leadership for the heavy emphasis on state control and investment in the Public sector. Hindsight does bring a lot of wisdom, and it is now obvious that countries have to evolve policies to suit the needs of a particular time, and specific situation. Would India be the kind of "industrial" and "knowledge" power that she is today, if it were not for the state driven initiatives of the 1950s and 1960s that created the IITs, the heavy industry, the research laboratories, the shipyards, the aircraft factories and many centres of excellence? On the other hand, would this country have realized its tremendous potential, had the barriers of state control not

Adapted from the inaugural address delivered at FICCI, at a seminar on "Public-Private Partnership in Defence Production, July 18, 2006, New Delhi. This seminar was organised by Naval Headquarters.

been brought down by Dr Manmohan Singh's historic initiative to liberalize and globalize in 1991.

So obviously, everything has a time and place, and it is clear that private participation in defence preparedness is an idea whose time has come.

The extent of private sector involvement vis-a-vis the defence outlay has been comparatively limited this far. Why is this so? To an extent this could be attributable to the acquisition procedures hitherto and non-involvement of the private sector at the project conceptualization stage. On the other hand, the private sector often looks at short-term investment and returns, which inhibit strategic investments. The inability to export is another constraint; since the quantities required may often be restricted, there have to be concerted efforts to promote exports, within the bounds of national security.

The first issue that needs to be addressed is: 'What' product is required, and secondly, the nature of participation of private sector. In this regard the Navy had prepared a 15-year indigenization plan that was well received by the industry. A Science and Technology roadmap has been drawn up for the Navy that identifies the 'end-product' capabilities that needs to be built over the next 20 years. This has recently been presented to the DRDO, and the Department of Defence Production, and we will pass it on to the industry too. This roadmap gives a clear picture of technologies and products that are foreseen for induction and will further help define what can be taken up by industry.

Ever since we launched the first Indian-built warship, INS *Nilgiri*, in the late 1960s, the shipbuilding industry has been the flag bearer of our drive for indigenisation. Today the Navy's force planning process is heavily dependant on the capacity and productivity of our public sector shipyards, and unless they can deliver a certain number of ships/ submarines a year, our force levels are going to slip. The good news is that the order books of the shipyards are now filling up rapidly, but the bad news is that with their best efforts, they lack the infrastructure, the capacity and the productivity to deliver ships at the rate that the Navy needs. So it is clear that the time has come to invite the private sector to contribute to warship building in whatever manner possible: public-private partnerships, joint ventures, outsourcing or subcontracting. Not only will this private-public teaming up create a most powerful synergy in warship building, but the resultant diversification will also be a force-multiplier for national security by creating centres of excellence.

The Navy's other focus area is Information and Communication Technology or ICT, in which the private sector has distinct strengths. The Navy is on the threshold of an exciting new era in which we hope to network all our platforms at sea; ships, submarines, aircraft, and Unmanned Aerial Vehicles (UAVs) with our operational centres ashore. To provide coverage over the entire Indian Ocean, we will need a ground segment and later a space segment. The Navy expects substantial participation by the private sector in building the structures for "Network Centric Warfare" or NCW, which is dominated by ICT. Providing connectivity for the Navy is a challenge, as dispersed forces operate over large ocean areas, making satellite and radio communication the only options. Moreover, the tri-axial movement of ships at sea complicates antenna design. The private sector has the strength for various building blocks in which software development will play a major role. The task is enormous and will have to be implemented in phases, as it must factor legacy systems already in service.

For all our sincerity and earnestness, the Navy's past endeavours to indigenize and to develop production sources in the private sector have lacked cohesion, continuity and above all, adequate financial support. We hope to overcome these handicaps through the creation of a Directorate of Indigenisation, which was inaugurated on the September 1, 2005.

I want to pay tribute to the wisdom and vision of my friend, Shekhar Dutt, who as the Secretary DP&S made a very generous offer to transfer a part of the Directorate General of Quality Assurance, which looked after indigenisation, to the Navy. With the approval of the Scorpene Project, we have embarked on the path of acquisition of national competence in submarine building. One of the first projects for this Directorate is going to be: to study the Scorpene submarine and to see how much of its components can be farmed out to private industry.

To move ahead, there is a clear need for dedicated groups comprising representatives from the Services, Department of Defence Production, DRDO and the Private Sector to address 'specific thrust areas' identified by each of the Services. These groups would be better able to define requirements, identify the model and extent of participation of the private sector, and work out the methodology for meshing in with current acquisition procedures and processes.

WHAT THE ARMED FORCES EXPECT FROM DRDO?

At the outset, I would like to say that there is a need to unequivocally highlight the importance of the DRDO in the Navy's scheme of things, and our sense of pride in your achievements. What we are proudest of is the symbiotic relationship that has evolved between our two organisations at all levels. This relationship has not come about as a matter of chance, but is something that both the DRDO and the Navy leadership have consciously and assiduously cultivated and nurtured over the years.

The fact that three of the DRDO labs carry the suffix "Naval" is only an external manifestation of this. Inside these labs, naval officers work shoulder to shoulder with scientists on developing cutting edge technologies, which have resulted in world-class systems for use at sea.

Not only that, we have also developed a process and a tradition of "putting our money where our projects are." Witness the LCA (Navy) project into which we have so far invested nearly Rs 500 crores. A naval aeronautical engineer is serving with the DRDO in a key role in its development. I do not think that there can be better synergy than that, between two organisations.

Today, we actually have at sea on our ships, and in the air on our aircraft and helicopters many systems developed by DRDO and produced by our PSUs, which are of world class. Amongst these are Sonars: hull-mounted, variable-depth and towed array; Radars: shipborne and airborne; Torpedoes; Electronic Warfare equipment: including ESM, ECM and ECCM suites; Missile and Torpedo counter-measures, Special Alloys, Hull Protection Systems, Special Paints and many other valuable operational innovations.

Much of what has been received is of excellent quality. There are some systems where, in spite of much labour and tremendous effort there are performance shortfalls.

Adapted from speech delivered at the DRDO's Annual Conference, Metcalf House, Delhi, February 10, 2006.

But we gratefully acknowledge the labours of our scientists, and would like to move on. So I have said that we will induct such systems into service as Mark I. But then we will demand Mark II, and that better meet our full requirements. The TAL torpedo is one such example.

What of the future? The way we look at it, we are fully committed to self-reliance and that means the future of the Navy is inextricably linked with that of the DRDO. It is worth briefly mentioning here of the Navy's force planning philosophy, whose foundation and corner stone is self-reliance. What does this mean for the Navy and DRDO?

- It means that we will build all the ships, submarines and if possible, aircraft and helicopters that we need, in India.

- Our current capability in warship building has some lacunae, especially at the high-end. We need to address them at the earliest to attain self-sufficiency.

- For example we have started making shipbuilding steel, steam and diesel engines, gearboxes, power generation machinery, hydraulic, pneumatic, air conditioning systems and some weapons. But there is plenty more that goes inside these ships.

- What we need to concentrate on now are advanced weapons, sensors, combat management systems and of course our ambitious networking programmes.

We have also decided that any 'transfer of technology' that we contract for in future will certainly not be the CKD, SKD or 'screw-driver' type that we have done so far, with no value addition to either our scientists or our industry. It will involve joint development, collaborative production and the right to develop (and market) a Mark II product.

In this context, I must acknowledge with gratitude the foresight and vision shown by the DRDO leadership in concluding the path-breaking tripartite collaboration with a foreign partner for the development of some advanced systems for our warships, which will be commissioned 2009 onwards.

In this project, we are sharing the funding as well as manpower liabilities with the DRDO. We are confident that this project, which involves DRDO scientists, naval engineers and industry representatives, will render tremendous benefits to the DRDO, the industry and to the Indian Navy. It should become a template for the future.

The first two items I was shown while at the DRDO stall at DEFEXPO on February 1, 2006 had exciting naval applications, fuel cells (for propulsion) and steel bulbars (for ship building). But I also read a sobering newspaper report, which stated that India today is the largest arms importer in the world. While we will always need to import some hardware, together we must ensure that not only do we reduce our import dependence, but that we become exporters as well.

What does the Navy expect from DRDO? That is to say what will lead to a better synergy with the Armed Forces.

Firstly, we see the DRDO as a "service provider" to the three Services. It must be pointed out however that the research carried out by DRDO labs cannot be an end in itself. It needs to be sharply focused at meeting some national defence requirement, whether it is in a long or short time frame. The faster the labours of the scientists in the lab bring a tangible benefit to the combatant in the field, the higher DRDO's credibility will rise.

I would like to give you an example from US operations in Iraq. They discovered that over 60 per cent of their casualties were from Improvised Explosive Devices (IEDs), mainly on roads, just as it happens to us in J&K. The US DoD constituted the Joint IED Defeat Task Force and tasked US Defence Advanced Research Projects Agency (DARPA), to come up with solutions. Within a year, implementation of the measures evolved by the Task Force has cut US casualties from IEDs by half.

There would be many such examples within the Indian Armed Forces too, where we need scientific solutions for combat related problems. For example, if IN ships have to be deployed in low intensity conflict operations in our neighbourhood, the biggest threat to them today would be from suicide bombers in high speed boats. It is a problem worthy of the DRDO, but do they have the inclination to address it? We are trying to induct electrical propulsion and air independent propulsion into our ships and submarines. Is it on DRDO's priority list?

Secondly, it is possible that DRDO does not come to know about Service priorities. I would suggest that DRDO scientists should visit field units of the three Services, on a regular basis to see the environment we function in, the problems our people face, and how DRDO can be of help. In fact I would recommend that a certain percentage of scientists should be recruited by the Armed Forces through the Short Service Commission route, and after five or seven years should transfer laterally to the DRDO.

Thirdly, frequent time and cost overruns of substantial proportions tend to erode the credibility of the DRDO in the eyes of the Services, which seek operational effectiveness to counter immediate threats. The solution however, does not lie in perpetual and unfounded 'optimism'. We must not shy away from facing the harsh reality of slippages, which are inevitable in research work. If you share your problems with us, the Services will not only understand your difficulties but also render necessary support. It is certainly worth considering collaborations and consultancies to overcome technological hurdles at an early stage of a project, which is languishing. A system of foreclosing projects, which do not fructify in a reasonable timeframe, should be strictly imposed.

Fourthly, increasingly, we are being asked for hardware from our maritime neighbours, but sadly, we do not have a ready response to make. DRDO and NHQ need to discuss, and put together a package of Indian designed and produced items like sonars, radars, communication and EW equipment and weapons, which we can offer to friends and allies in the neighbourhood.

Fifthly, one of the biggest problems that we face on ships is the eternal struggle between the laboratories and the production agencies, when a newly installed indigenous system malfunctions on a ship. I think that it would be well worth investing in a special test site and even a research ship, where complete integration of a system could be jointly undertaken as part of ToT between DRDO and the production agency before installing it on a ship.

And finally, an unsolicited suggestion. We know that winds of change are blowing in our country, and one day possibly DRDO may also be asked to reinvent itself. In anticipation, it would be prudent to study defence research organisations elsewhere.

The DARPA (the inventor of the Internet), owns neither scientists nor labs, but hires bright young people from industry to undertake research projects. In Israel, defence research has been merged with the production agencies and this synergy generates both speed and efficiency in successful materialization of projects. The UK, in 2001 converted their Defence Evaluation & Research Agency (DERA) into a private-public enterprise and renamed it QinetiQ. Today the MoD holds 56 per cent shares in QinetiQ, which carries out research for the defence, security and commercial sectors and is a profit making company listed on the London Stock Exchange. Perhaps some similar systemic change may help re-energise and improve the performance of DRDO. Once you begin to generate profits, you will be able to retain your young scientists.

Role of Armed Forces in Disaster Management

Ever since the beginning of history, people have been worrying about the "end of history". Most major religions tend to predict the demise of the world, and it is known variously as Apocalypse, Armageddon, and *Qayyamat,* or *Kalyug* in this part of the world. The medieval writer Nostradamus seems to have added to "gloom about doom" by even predicting a date, which for your peace of mind, I shall refrain from mentioning! The frequency and intensity of natural calamities of the past year seem to have encouraged speculation about the impending demise of our planet, and this is not just amongst the superstitious. Fortunately, we are discussing here issues, which are a little more rational.

Events of the recent past have brought an intense focus on the need for a ready response to disasters, which as I mentioned, have increased, both in their frequency and scale. This can be ascribed substantially to climate change, for which humanity has only itself to blame. Another reason is the growth in urban population, because of which people have started settling in coastal, low-lying or reclaimed areas, which were earlier either underwater or prone to flooding during heavy rains. This disorganized urban growth is underpinned by economic factors and little can be done about it.

Whatever be the reasons, and whether one believes in Nostradamus or not, it is clear that disasters are going to be more frequent with larger number of people being involved. Governments and citizens the world over look up to the Armed Forces as the first agency to come to their assistance in any large-scale crisis. The reason for this is self-evident; armed forces constitute the biggest national organisation for providing rescue and relief, and are trained to respond swiftly to disasters and emergencies of all kinds; be they natural or man-made.

Adapted from a Keynote address at an international seminar on Disaster Management, organised by the HQ, Integrated Defence Staff and the UNOCHA, at the United Services Institution of India, New Delhi, October 28, 2005.

In 2004-05, for example, India was struck in rapid succession, by a tsunami, floods, heavy snowfall and an earthquake. The resultant situations required the use of a combination of equipment and facilities: ships and helicopters of the navy, heavy airlift capability of the air force, and the engineers, and ubiquitous infantrymen of the army. All three Services contributed significantly in providing medical assistance from a common pool.

Apart from the above attributes, the training, discipline and organisation of the Armed Forces enables them to carry on even in the face of extreme adversity, where other organisations seem to panic and lose cohesion. It was notable that during the October 2005 earthquake in J&K, despite suffering casualties within their own ranks, the Indian Army deployed rapidly to help the affected civilian population without breaking stride.

Each disaster episode throws up its own lessons for the future. But I will just flag a few salient issues for your consideration here.

The first one is that of readiness. While ships and aircraft are typically capable of being launched in a matter of hours, this is of little avail if the relief material and the manpower needed for the actual relief effort, are not available within the same timeframe. In disaster situations, time is the essence if casualties are to be reduced and the relief operation is to be effective. Consequently, we need to look at "composite task forces" for disaster relief in which ships, aircraft, manpower, equipment, relief material and even personnel are earmarked and kept ready for a variety of possible disaster situations. Standard Operating Procedures (SOP) need to be prepared so that old lessons are not re-learnt after every disaster.

The second issue is that of Command and Control. This can impact on sensitivities, not only between different nations, but also between different services and agencies within a nation. Unless interagency lines are clearly drawn and boundaries demarcated, there is potential for conflict and confusion. This is an area that should receive sufficient advance attention from planners, because such friction has the potential to debilitate and derail relief operations. Disaster relief operations are the ideal nursery for "Jointmanship". In our own case, a very good model was demonstrated by the HQ IDS, which conducted the tsunami relief operation with great efficiency.

The third issue is the suitability of military equipment and hardware for disaster relief. For example, in the aftermath of the tsunami, our Navy found that many of its ships were not ideally suited for operating in an environment where most of the jetties had been washed away and existing beaches had been submerged. In such a situation, there was a pressing need for air cushion vehicles and heavy lift helicopters to access remote islands and coasts. The Services need to take a fresh look at how we draw up the qualitative equipment requirements, and train our units so that they can easily be reconfigured or reenrolled for disaster relief operations. Dedicated training capsules for disaster relief operations is also a connected issue that needs attention.

The last issue to be flagged is that of coordination with civil agencies, international organisations and NGOs. Very often, the Armed Forces tend to think that they are the beginning and end of disaster relief operations. The fact is that while very often we may be the first on the scene, the civil administration, agencies like the United Nations and NGOs continue to work months and sometimes, years after we depart. Often there is also duplication of effort, and lack of coordination, which render these operations much less effective. The remedy for this is obviously to bring synergy into the planning and execution of relief operations right from their initiation. Umbrella bodies like India's newly created National Disaster Management Authority may be the appropriate agency to coordinate and synergize the endeavours of all those who need to contribute to disaster relief.

Looking at the future with optimism, we in the Armed Forces sincerely hope for a reduction in the scale and frequency of human conflicts, worldwide. But it is unlikely that we would be either willing, or able to "beat our swords into ploughshares" in the near future. At the same time, it is obvious that mankind will remain vulnerable to the vagaries of nature, which will continue to wreak havoc and misery on civil populations. What better use can there be for our skills and expertise than to render succour to suffering humanity?

SPECIAL FORCES IN THE INDIAN CONTEXT

I do not claim to have any first hand knowledge of the Special Forces, except for a brief encounter that I had a couple of decades ago. In 1987, while in command of a frigate, I was ordered to embark a detachment of 10 Para from another ship at sea and proceed to anchor off Colombo. The mission was to keep an eye on the Presidential Palace and render help by helicopter in case things went out of control after the signing of Indo-Sri Lanka accord. We lay at anchor just off Colombo, while the Para Commandos, masquerading in naval uniforms, kept continuous vigil. However, our hopes of high drama were belied, and after about a week we were sorry to see the Paras leave us without seeing action.

The subject of Special Forces has both relevance and importance in today's security environment, and this will be even more so in times to come. The battlefield scenario has undergone rapid change in the past few years and combat will now take place between highly mobile and networked forces, operating in a transparent battle-space with long-range precision-guided munitions at their disposal. Warfare, as we have known it, may soon be a thing of the past because technology, rather than manpower has now become the determinant of battles. There are already indications that even the conventional battlefield may no longer exist, because conflicts will now take place on different planes altogether.

If our adversaries of tomorrow, or their proxies are going to be non-state actors; terrorists, insurgents, pirates and hijackers operating at a sub-conventional level, then our response too, will have to be at the same level. In such a scenario, Special Forces will certainly play a most significant role, and perhaps even take centre stage.

We know from bitter experience of the past 60 years that India cannot afford to let its guard down against external threats or internal subversion. Moreover, as a growing

Adapted from the introductory address delivered at a seminar on Special Forces organized by the Centre for Land Warfare Studies, New Delhi, March 13, 2006.

economic and military power, we need to abandon our traditional inward-looking attitude, and turn our gaze at the external strategic environment. Then we will realize that our geographical area of interest and the concomitant responsibilities have indeed become vast and widespread.

Security challenges of the future are going to be more ambiguous, and more complex, and will require multifaceted responses. We also need to accept that our adversaries are going to relentlessly try to erode India's power and economic status by posing non-conventional threats and by waging asymmetric warfare against us. We do not seem to realize this; but what we have been facing on a daily basis for a decade and a half is actually sustained asymmetric warfare. And Special Forces have a role to play in it, on both sides.

We also need to recognise that the response to challenges of such a nature do not always lie in the military domain. And here it is germane to recall the oft quoted words of Clausewitz: "War has no autonomous existence, except as an instrument of policy, and every act of war must have a clear political objective." Regrettably, in the past we have not often paid heed to these words, either in the prosecution of our wars or in their termination. However, where asymmetric warfare is concerned, our actions must have a long political underpinning.

Ideally speaking, Special Forces should be capable of undertaking special operations at the strategic, operational or tactical levels, a key role being to provide information to assist decision-making at the strategic and operational levels. In India perhaps, we have tended to treat Special Forces as an adjunct to conventional troops, and therefore focused them at a tactical level to help attain battlefield objectives. Other countries employ Special Forces to apply calibrated pressure at precisely calculated points to achieve political effect, and not merely battlefield victories.

The two most famous Special Forces actions in WW II were both at a strategic level and had far reaching consequences. The first one was the rescue of Mussolini from captivity by elite German commandoes in 1943, to prevent Italy from capitulating to the Allies. The second one was the Allied attack and destruction of the Norwegian heavy water plant under German control in Telemark in 1944 so that the Germans could not produce an atomic weapon.

In India we have today, between the three Services, the Home Ministry and Cabinet Secretariat, Special Forces, which are in excess of 10,000 strong. This constitutes a very significant national asset, and the organisation of the structures; manpower, equipment and training of Special Forces are issues, which merit serious consideration.

While we may have over the years developed our own perceptions and concepts on deployment of Special Forces, we need to keep our minds open and look closely at the way others do things. We must, therefore, examine the doctrine and philosophy of countries that have highly professional Special Force units; whether it is the Green Berets, the SEALs, the SAS, or the Russian Spetsnaz.

For us, a subject worthy of study would be the role played by the Pakistan Special Services Group (SSG). This group, amongst its other activities, has been employed in the 1950s in the Naga Hills, in 1971 in the erstwhile East Pakistan, in the 1980s in Afghanistan and of course ever since its inception, in J&K. The SSG has all along been tasked for objectives, which are certainly on a level much higher than the tactical. Over the years, four battalions of the SSG appear to have achieved results disproportionate to their strength.

The US Quadrennial Defence Review is available on the Internet, and the section dealing with Special Forces makes interesting reading. While acknowledging the vital contribution of Special Forces during Operations Enduring Freedom and Iraqi Freedom, the Review recommends the enhancement of manpower, acquisition of new technologies, and procurement of new platforms for the US SF. Amongst the salient roles, missions and recommendations are:

- Conduct of unconventional warfare in dozens of countries simultaneously.

- Greater capacity to detect, locate and render safe WMDs.

- For direct action, the ability to locate, tag and track dangerous individuals globally.

- It is envisaged that SFs will build language and cultural skills specific to key areas in Middle East, Asia, Africa and Latin America.

- Active duty SF battalions will go up in strength by 33 per cent. It is planned to create a separate Marine Corps Special Operations Command.

- A SF UAV squadron is to be created, and conversion of four ballistic missile submarines into special operations platforms undertaken.

This is just a glimpse into the US thought process regarding their Special Forces, and could provide some good road signs for us.

In the emerging strategic scenario, we in India need to take note of the words of Clausewitz that I just quoted, while undertaking a review of the structuring and organisation of our various Special Forces. Demarcations need to be drawn between conventional, airborne and Special Forces. We must consider taking our Special Forces partially or completely out of the tactical domain so that they can operate across the full spectrum of conflict in non-conventional roles.

And finally, it should be our endeavour to place the Special Forces in an organisational hierarchy and framework such that their employment makes an impact on the affairs of State at the strategic level or even at the grand strategic level.

THE MARITIME DIMENSION OF NATIONAL SECURITY

I am indeed privileged to be present here today to deliver this year's General B.C. Joshi Memorial Lecture, which has been arranged by Pune University to mark this distinguished soldier's tenth death anniversary. Those who had the privilege of knowing or serving with General Joshi are well aware that he was a farsighted person; and one of his fondest visions was to ensure the spread of learning at all levels of the Indian Army. He was particularly keen that army officers should spend time in academic institutions, and had therefore, established a unique and rather close relationship with academia, particularly with the Pune University.

General Joshi was instrumental in the setting up of the Chattrapati Shivaji Chair in Policy Studies at this University, with the aim of encouraging research into National Security issues. I think this was the first such chair instituted in India, and by a coincidence, just last month I happened to inaugurate a Chair in Maritime Studies at the Calicut University. I feel that in a sense, the Navy was just walking down a trail, which had been pioneered by General Joshi.

Addressing the All India Vice Chancellor's Conference at the Pune University on, November 13, 1994 barely five days before he passed away, General Joshi had stated, "The next century will be India's century. This is not merely wishful thinking or building castles in the air, but the considered opinion of a soldier who has studied events and developments carefully." As India marches ahead into the 21st century Gen Joshi's words are proving to be prophetic indeed, and stand as testimony to this fine soldier's vision.

"War", it has been said "is too serious a business to be left to the Generals." In developed countries, almost all research in the field of national security is carried out in

---

Adapted from address delivered at the Annual General B.C. Joshi Memorial Lecture organized by the Pune University, Pune, November 11, 2004.

think tanks and institutions of higher learning, with the participation of academics as well as retired and serving Armed Forces officers. A similar trend is emerging in India too, and one finds that industrial houses and universities are willing to invest intellectual as well as financial resources in researching national security issues. The national security matrix today, encompasses subjects ranging from economy to environment and from weapons to water. While this trend is indeed welcome, we as a nation do have a somewhat flawed perception of national security.

Flawed Security Perceptions

Such is our preoccupation with the politics of the day that sadly, national security matters generally get relegated to the background. Attention on such issues tends to get focused only during periods of a crisis, and the minute it is resolved, the nation seems to lose interest. There can be no doubt that the fundamental quest of our nation state should be to provide a better quality of life to its teeming millions. However, the argument between guns and butter is not simply a zero sum game. If the past fifty-seven years have shown us anything, it is that you just cannot have development for your people unless you create a secure environment for the country. And a secure environment will be achieved only when it is clear to everyone in the neighbourhood that you are a strong confident nation with a surplus of security assets and the resolute will to act in your national interest.

It would not be an exaggeration to state that our detached attitude towards national security (which includes intelligence) has been the reason for the imposition of various conflicts on India by our neighbours from the invasion of Kashmir in 1947 to the occupation of the Kargil heights in 1999.

There are two major shortcomings in our system, which have a bearing on this state of affairs. The first of these is our ambivalent attitude to national security and defence expenditure. Every few years we tend to agonize over questions like: is there a threat to our national security? Do we need to commit funds to the Armed Forces on a long term? This attitude has led to fluctuating budgets and uncertainty in the force planning process. To compound the problem there is a total mismatch between the life of the defence budget, which is about 9-10 months and our decision-making process, which takes a minimum of 2-3 years. It should surprise no one that year after year, we are unable to expend the money voted for defence by parliament. The present system of defence acquisition is guaranteed to ensure this.

This brings me to the second factor; a strategic vision, or rather, the lack of it. The question whether we as a nation and a cultural entity have lacked a tradition of strategic thinking and planning has been discussed often in recent times. While this debate may go on, the fact remains that for a variety of reasons, at present, our system does not take a long-term view of national security and that is why we often tend to react in a knee-jerk and uncoordinated manner to emerging threats and realities.

Every Indian citizen is a stakeholder in the nation's security. It is, therefore, necessary for not just the political establishment, but also the media, the bureaucracy, and the academia to engage in a meaningful discourse on substantive issues. This would not only provide focus and a sense of direction to such debate, but would also remove the perception that India is a soft state.

If indifference to security issues is a national malaise, then disinterest in maritime security issues is a tendency, which is of equal concern. It is said that those who do not learn from history are condemned to live through it again. So with this in mind, let us delve a little into our maritime past.

The Indian Ocean and India's Maritime Past

"He who rules on the sea will shortly rule on the land also," was the advice given by the 16th century Ottoman Admiral Khaireddin Pasha (aka Barbarosa) to his Sultan Suleiman 'the Magnificent' of Turkey. The history of no country illustrates this principle better than that of India. A majority of the invasions of India were from the landward side. But such invasions and conquests either led to transient political changes or to the foundation of new dynasties, which in a very short time were absorbed by the resilient fabric of India's culture, and became Indian. In fact, it may truly be said that India never lost her Independence till she lost the command of the seas.

In ancient India, while the West Coast witnessed mainly commercial activity, the eastern waters were used as a medium for establishing longer lasting relationships. The deep cultural linkages of the subcontinent that are still very much in evidence with Myanmar; the Malayan peninsula, the islands of Sumatra, Java, and Bali in Indonesia, and with Thailand, Laos and Cambodia, stand witness to this.

The prowess at sea shown by the Mauryan dynasty, to which Emperor Ashoka belonged, was followed by maritime forays of the Pallava and Chola dynasties that set up surrogate empires in (South-East) Asia. Indian seamen of that time were also at the forefront of developments in seamanship and navigation, and the invention of the magnetic compass or "Matsya Yantra" as it was then called, is ascribed to Indian navigators in ancient Egyptian texts.

I bring up these facts before this audience because most Indians, much less foreigners, are unaware that they are heirs to a rich maritime heritage, that their ancient forebears knew the seas well and used them for commercial, political and cultural purposes. Hence we are neither strangers to the sea, nor interlopers in the Indian Ocean.

Till the beginning of the 16th century, because of the vast size of India and the insulation that Indian maritime power provided, oceanic problems had not really intruded into the history of the mainland. In fact, till the arrival of Portuguese at Calicut in 1498, no naval power had appeared in Indian waters. The arrival of Vasco da Gama changed all that. From that time onwards the powers that dominated the oceans also dominated India.

The glory of the Mughal Empire on land could not hide their bankruptcy at sea. The Mughals with their Central Asian tradition had unfortunately, no recognition of the importance of the sea. The result was that during the two centuries of the Mughal Empire, not only did the Indian Ocean fall entirely under alien control but, simultaneously, the foundation was laid by others for a more complete subjugation of India than any land power at any time could have conceived.

The decline of India's sea power by the 14th century in the Mughal era was to a large extent responsible for the success of the European adventurers who began to arrive on our shores in the next century. The Portuguese came first, followed by the Dutch, British and the French. They came in search of spices, but stayed on to rule this land. The importance of the sea was realized by Indian rulers only when it was too late. In fact, apart from a few visionaries like Kunjali Marakkar and Kanhoji Angre, the Indian maritime scene was bereft of both the intellect and the will to use the seas to our advantage.

With Robert Clive's victory at the Battle of Plassey in June 1757, the Indian province of Bengal fell to the British. The thin end of the wedge was in position, and this date is commonly accepted as the beginning of Britain's 190-year rule of India. Commenting upon this historic event, Admiral Mahan, in his seminal work, "Influence of Sea Power on History" remarks, "...it may be said that the foundation thus laid could never have been built upon, had the English nation not controlled the sea. The conditions in India were such that Europeans of nerve and shrewdness, dividing that they may conquer, and making judicious alliances, were able to hold their own against overwhelming odds."

That our country has been prey to centuries of invasion and conquest, and that final domination by an alien power resulted not by overland invasion but by an onslaught across the seas is a fact that should be indelibly engraved in the mind of every Indian. It should also form the foundation and underpinning for our maritime thought. The United States succeeded Britain as the world's pre-eminent maritime power in the mid-20th century and remains so to this day. It is as a direct consequence of this attribute, that the US is also the pre-eminent power in the world today.

Mackinder and the Heartland Theory

Most of us look at history as a record of inevitable events in human affairs, but there are those who study it against the backdrop of realpolitik and look for cause and effect. One such person was Sir Halford John Mackinder, a great scholar of political geography.

Examining the status of sea power in a geo-political context, Mackinder postulated the view that world history was a cyclic struggle between land-based and sea-based powers. He cited the example of Europe, where until 1500 AD East European land powers from the European "heartland" had dominated the continent. However, the period from 1500-1900 AD, which he termed as the Columbian era, witnessed the rise of "rimland" maritime powers like Britain, Spain, Holland and Portugal, because they exploited the mobility provided by the sea to give them an advantage over their territorial adversaries.

What Mackinder foresaw in 1904 was that the traditional advantage of mobility enjoyed by sea power was now being met in equal measure by mobility on land. The internal combustion engine, and the consequent development of the road and railway systems, had wrought this significant change. In 1919, he articulated the famous dictum that Eastern Europe was the key to the Heartland of Eurasia, which in turn commanded the World-Island. The power, which ruled the World-Island, ruled the World. It was this thesis that made Hitler roll his Panzers into Soviet Russia in 1941 in a desperate bid to gain control of the Heartland, prior to his intended conquest of the world.

Mackinder warned that the great continental powers in the heartland, with their hinterland full of resources were ascendant. They would soon develop their capabilities to a point where they would break the barriers imposed upon by the rimland and surge out to the sea. In a sense he foresaw the demise of Great Britain as a world power half a century before it actually happened, and till the late 1980s, when the erstwhile Soviet Union and the US were the two superpowers, it was fashionable to quote Mackinder's prophesies.

What people failed to notice was the fact that the US, which was by then of course a continental power, had also developed into a maritime power, whereas the Soviet Union remained essentially a continental power. One analysis sums up the developments of the past 25 years as follows: "The United States actually encircled the Soviet Union under the famous Doctrine of Containment, principally using its maritime power, principally aircraft carriers and nuclear submarines, as the vehicle for this effort. By enlisting the rimland countries against the Soviets, by using Central Europe as a bridgehead, Great Britain as an advanced air base, and Canada and USA as resource bases for material and manpower, they denied the Soviets access to the sea and ultimately won the Cold War."

Some may dismiss this as a facile and simplistic explanation of a momentous historical event beset with complex geo-strategic and economic nuances. However, an analysis of the last few centuries does show that continental powers have historically fared poorly against maritime powers. The reason is simple. The seas offered unrestricted space for movement and access, which when exploited led to immense economic gains. This in turn, gave countries the ability to finance a large military, including a powerful navy, and thus the cycle goes on.

Today the USA, the dominant maritime power of the day, is also the sole superpower. Russia, whose status as a first class power is now questionable, has to vie with the USA for influence over Central Asia, which is an enormous repository of fossil fuel. Geo-politics is however an uncertain science, and whether the future will mirror the past remains to be seen.

We now turn to India's stake in the seas that surround us, or what is termed as, our maritime interests.

India's Maritime Interests

Pandit Jawaharlal Nehru had once stated: "...We cannot afford to be weak at sea. History has shown that whatever power controls the Indian Ocean, has in the first instance, India's seaborne trade at her mercy, and in the second, India's very independence itself." India's national objectives lie in ensuring a secure and stable environment, which will permit the economic development and social uplift of its masses, so that the country can ultimately take its rightful place in the international comity of nations. Within this overall objective, our prime maritime interest is to ensure national security and provide insulation from external interference, so that the vital tasks of fostering economic growth, and undertaking developmental activities, can take place in a secure environment. In this context, it is worth mentioning here that national security also encompasses certain economic factors in the maritime domain.

Overseas Trade. Reforms in the past decade have resulted in a liberalised Indian economy, integrated with, and increasingly interdependent on other world economies. India is now projected to become the fourth largest economy in the world by 2020, after China, Japan and the US. Our foreign trade of approximately US$ 95 billion accounts for about a fifth of our GDP. Of this trade, 77 per cent by value and over 90 per cent by volume is carried by sea. The tremendous scope for further growth can be imagined when we consider that our share of the world trade is only 0.62 per cent.

Energy Needs. The critical importance of a continuous and assured supply of energy resources such as oil, natural gas and coal for a growing economy like ours cannot be underestimated. According to current projections, with steady industrial growth India would, in a few years become the world's largest consumer of petroleum products. With limited domestic production, the import element, currently at 60-70 per cent is going to steadily keep rising. While talk of pipelines continues, we may safely assume that ships will carry much of these energy imports across vulnerable sea-lanes.

Undersea Resources. Another facet of the ocean, which presents the prospect of wealth and prosperity, and yet contains the seeds of future conflict, are undersea resources. Our offshore hydrocarbons represent a vital asset as well as a maritime liability. The average depth of the Indian Ocean is about 4 km and that is the distance, which separates us from a treasure house of rare minerals, in the form of poly-metallic nodules awaiting exploitation on the ocean bed. India has a mineral rich Exclusive Economic Zone (EEZ) currently extending over 2.2 m sq km, and the successful exploitation of these could lift us from economic backwardness.

Merchant Fleet and Ports. India's merchant navy is small for our needs, and size. Relatively speaking, it constitutes a little over one per cent of the world shipping tonnage, and our ships are able to carry only about a third of our own foreign trade. However, in absolute terms, India's growing fleet of over 600 ships is quite large, and operates out of 12 major and 184 minor ports. The security of these ports, our merchant ships and the sea-lanes that they ply on represent vital maritime interests for us.

Attributes of Maritime Power

In 1905, when the Royal Navy was seriously contemplating drastic cuts in the size of the fleet as an economy measure, the British Foreign Office complained bitterly: "If the number of ships is to be reduced to an extent that the Royal Navy will not be able to give the foreign policy of this country, such support as is expected by the foreign office, we can only conclude that the vital interests of the country are being sacrificed in the interest of false economy."

This was, of course, a statement made in the heyday of what was known as "gun boat diplomacy." But today, more than ever, navies form a potent instrument of state policy, provided wisdom and imagination exists, to wield this instrument in furtherance of national interests. The oceans remain the "common heritage" of mankind and freedom of navigation on the high seas is a fundamental principle of maritime law. Unlike every inch of land on the continents, the seas in their natural state are free of political control.

Seventy per cent of the earth's surface is covered by the sea and over 2/3rd of the world's population lives within 100 miles of it. Over 150 of the 185 member states of the UN are coastal states. Since the 1970s they have all extended their jurisdiction out to sea - in many cases, as much as 200 nm and some beyond that. Most human maritime activity is confined to within 100 nm of the coastal zones. This means that a very large proportion of the world's economic, industrial and political activity is carried out in this swath of land and sea extending about 300 nm. This is often referred to as the littoral, and that is where the navies of the major maritime powers are focusing today. It is noteworthy that in all of the recent major conflicts - Kosovo, Afghanistan and Iraq - the major intervention by coalition forces has not been overland, but from the littoral. The intervention came through naval aviation, through launch of cruise missiles by ship or submarine or by landing of marine expeditionary units.

All applications of sea power flow from the fact that the sea is the greatest of highways. It is much easier to move a ton of commodities, oil, or military hardware by water than by land or air. Contrary to Mackinder's beliefs, the US finds it cheaper to ship semi-conductors from Malaysia or cars from Japan today than to carry the same items from east coast to west coast by land or rail. The fundamental quest in maritime warfare, therefore, has always been to secure the free use of the sea highway for oneself and to deny it to the enemy because the economic lifelines of nations run across the seas.

It is in this context that I would like you to look at certain distinctive attributes of maritime power, which make it unique and invest it with the potential to act as the handmaiden of state power or diplomacy without even leaving a footprint in its wake.

- **Access.** The sea covers over $2/3^{rd}$ of the globe and this allows maritime forces to exploit the oceans for their unhindered deployment in areas of interest or threat. International law provides free and legal access for ships up to the territorial waters of nations, and also allows the right of what is called innocent passage for traversing these waters. As Operation Enduring Freedom showed recently, even landlocked states can become accessible from the sea provided intervening coastal states cooperate. At the same time, the medium of the sea provides the contiguity to convert distant nations into neighbours - friendly or otherwise. The UK and Argentina, separated by 2,000 miles fought a war through the medium of the sea in 1982.

- **Versatility.** Warships can easily change their posture, undertake several tasks concurrently, and be rapidly available for re-tasking. Warships at readiness are always manned and provisioned to sail at short notice for ops commitments. Individual ships of frigate size and above will always have capabilities in all three dimensions, and warships can be formed into task forces, which provide a mutually supportive combination of offensive and defensive capabilities.

- **Sustained Reach.** Maritime forces have integral logistic support including refuelling, repair and medical facilities. The range and endurance that these provide gives maritime units sustained reach, which is the ability to operate at considerable distance from shore support.

Role of the Indian Navy

It is seen against this backdrop that maritime forces have an application in a wide range of operations at sea, extending from nuclear conflict or high intensity war fighting at one end of the spectrum, to humanitarian relief and stable peace at the other. This range of operations can be broken down into four compartments or types of roles, which may in general terms, be described as military, diplomatic, constabulary and benign:

- **Military:** Where combat is used or threatened, or which pre-supposes a combat capability. This includes the application of maritime power at sea, and also from the sea in both offensive operations conducted against enemy forces, and defensive operations conducted to protect friendly forces and maritime trade. In our context, Sea Control, Sea Denial, Blockade, and Low Intensity Maritime Operations (LIMO), are some examples of this role.

- **Diplomatic:** Military force at sea contributes to what has traditionally been known as Naval Diplomacy. It is the use of maritime forces as a diplomatic instrument in support of political objectives and foreign policy. It assumes the availability of force to back up and support diplomatic efforts at various levels.

- **Constabulary:** Where forces are employed to enforce law, or some regime established by international mandate. In this role, violence is used only in self-defence or as a measure of last resort in execution of the task. In the Indian context, some of these tasks have been assigned to the Coast Guard.

- **Benign:** Tasks such as humanitarian aid, disaster relief, Search and Rescue (SAR), salvage assistance or hydrography are classified under the heading of "benign". Such tasks may require the specific capabilities and specialist knowledge of our forces in an emergency. Violence has no part to play in its execution, nor is the potential to apply force a pre-requisite.

In general, the Indian Navy can be the catalyst for peace, and stability in the Indian Ocean Region, across a wide range of conditions and circumstances. The imaginative use of the navy can achieve this in more ways than one. It can be used to engage other maritime nations and extend our hand of friendship and cooperation. Its robust presence in a particular area or region could contribute to stability and ensure peace. Lastly, it can act as a strong deterrent with the ability to prevent conflict, or to respond, should it become necessary.

FUTURE STRATEGIC CONTEXT

It is an accepted tenet of international relations, that nations have no permanent friends or enemies, only permanent interests. The order of battle of the IN is not configured on the basis of perceived threats to our security, but on the basis of capabilities existing in our neighbourhood and the likelihood of emergent challenges. The underlying premise is that if a capability exists, or is being developed by a country with which we share boundaries or interests, it could have a bearing on our security, should circumstances or intentions change over time.

The Security Scenario

To start with, our field of view extends from the Persian Gulf, down the East coast of Africa, and across to the Malacca Strait, which contains the area of direct interest to us. Let us look at who are likely to be the major players and what are the issues in this region? Certain international security trends can be discerned based on current form.

The US will certainly remain the sole superpower for the foreseeable future. The EU, China, Japan, Russia and India are likely to be the other reckonable powers in the world. The US will remain actively engaged in the Asia-Pacific region. Underpinning all her policies is the ever present and overriding need to have access to the oilfields and markets of this region, so that her own oil reserves remain intact, and her economy buoyant. The US also has a requirement to fulfil security obligations to countries like Japan, Taiwan and South Korea. A concomitant requirement is to contain or engage China's growing power. It is interesting to see a gradual transition of US security policies to suit their national interests, from a Cold War posture to her present stance.

Therefore, as Prime Minister Manmohan Singh stated, "For India, engaging the US is a necessity." The Indian Navy has been a prime instrument for this engagement and many may not know that the first military engagements with the US after the end of the Cold War were in the form of naval exercises that were first carried out in 1991. The reason why the Navy was the forerunner in breaking ice with the US was of course the fact that ships and fleets operate beyond the horizon, and leave no "footprints" on land. Since then we have progressed to the stage of advanced level exercises where even their nuclear submarines participate regularly.

The fountainhead of terrorism and focal point of nuclear and missile technology proliferation is located right next door to us. Defying predictions, our neighbour Pakistan has made a remarkable economic recovery, and by playing her cards right, derived maximum advantage in the post-9/11 scenario. From recent experience, she has learnt a cautionary lesson about maritime security, and with substantive help from China, and perhaps USA, she is in the process of adding hardware to her navy. China is also constructing a strategic alternative to Karachi port in Gwadar on the Makaran coast. Gwadar sits astride the exit to the Persian Gulf, and a Chinese naval presence there would be of concern to us.

China's defence spending today runs at about five times ours. She is taking full advantage of the prevailing tranquility, and her buoyant economy, to modernize her armed forces. Even if we discount any expansionist designs, China's desire to safeguard her growing energy dependence on the Gulf will see her building up the PLA (Navy). While major warships, diesel submarines, and long-range strike aircraft are being sourced from Russia, her own industry is producing nuclear submarines and long range "ballistic missiles. China is intensely engaged with almost all countries in our immediate neighbourhood. She has in place, agreements for supply of military hardware to these countries, and maintains very cordial relations with them. Which is more than we can say for ourselves.

It is an unfortunate fact that for various reasons, our relations with countries all round are under a degree of strain. In my opinion, we have for far too long, neglected our smaller neighbours and demanded "reciprocity", when our relative size, economy, and above all tenets of good-neighbourliness actually suggested "generosity".

Trade interdependence within the region is already a reality and will grow even further. As a result of the economic boom, the Asia-Pacific region's share of world energy consumption will continue to rise and energy security will assume progressively greater significance. India's own oil consumption is expected to rise to 150 million tons by 2020, with the country likely to become the world's single largest importer of oil by 2050. Consequently, stakes in security of the vital sea lanes and the pressure for countries to cooperate in the maritime domain will grow enormously. Any challenge to the free flow of oil, which can be interrupted by a host of state and non-state agencies, can lead to major conflict, which can have profound effects on the world economy.

Terrorism will continue to unite and divide nations. Terrorist threats of transporting men and material (especially WMD) through sea routes will continue to drive multilateral security initiatives by the US. Three of them - the International Ship and Port Security Code (ISPS Code), the Proliferation Security Initiative (PSI) and the Container Security Initiative (CSI) are already in various stages of implementation. Once again, the maritime dimension is taking the lead in implementing these collective security measures.

LIMO involving political as well as criminal entities, will gain in importance. The Indian Navy and the Coast Guard have been engaged on security duties in the Palk Bay, along our maritime boundary with Sri Lanka, since the late eighties.

On the West coast too, the Navy has been conducting surveillance operations since 1993, consequent to the Bombay blasts. With the complete fencing of our western land borders, a resurgence of clandestine activity at sea can be anticipated.

Conclusion

Let me conclude by leaving some food for thought. When I interact with my counterparts in other navies, especially some from the Western hemisphere, the questions most frequently put to me are: What is the purpose of your large navy? Why are you buying or building ships and aircraft carriers? Why does your Maritime Doctrine talk of a sea based nuclear deterrent?

From this line of questioning, it becomes obvious that the colonial mindset does linger in some quarters. In the old days, it was understood that maintaining peace, law and order east of the Suez was the "white man's burden", and it still appears preposterous to some that a third world navy should presume to shoulder it. The best that can be done is to put across a balanced Indian perspective on such matters, and hope that one's interlocutor is convinced.

But, what does one do when, the same or similar reservations are often expressed by one's own compatriots? Essentially the sceptics say that all our security problems are concentrated in the north and east, so why spend money on a Navy? Of late, defence expenditure is also being questioned in the context of peace talks, and CBMs under discussion.

I would like to draw attention to the fact that historically in international relations, a power imbalance in any region has always led to instability and eventually to war. In the Asia-Pacific region, there are a number of economic and military power centers, either existing or emerging, and the pre-eminent among these are China and India. It is in the nature of things that the stronger power will try to dominate and establish hegemony over the others.

In such a situation, unless a balance of power can be established, either the weaker power capitulates without struggle, or there is a conflict. History has shown that appeasement only whets the appetite of totalitarian states. While we have no wish to dominate anyone, we need to ensure that nobody else is in a position to dictate terms to us either. And for this, it is essential to build and maintain a credible deterrent military capability of which a strong navy is an essential ingredient.

With the global security environment now focused on Asia-Pacific, the Indian Ocean Region has become critically important to many major powers. Today Asia-Pacific region contains almost 4 billion of the world's over 6 billion people and accounts for 60 per cent of the world's GDP. By 2020, seven of the ten largest economies in the world will be in this region, making the 21st century truly the "Asia-Pacific Century". While the region holds great economic potential, it also contains the seeds of conflict in several areas, and has lately become prey to the scourge of religious fundamentalism.

For India, the Asia-Pacific region holds immense promise for political, economic and military cooperation, and the key role that maritime forces can play, makes the Indian Navy an important component of any national strategy designed for this region.

However, a capable navy is only one element of maritime power. A large merchant fleet, modern ports with good infrastructure, a vibrant, efficient and self-reliant shipbuilding industry along with its supporting technical infrastructure are some of the other vital ingredients of maritime power which we need to concentrate upon.

The key, however, lies in the populace having a maritime temperament and outlook. Indians in general, need to acknowledge that we certainly are a nation dependant on the seas, and need to look more seawards than inwards. Such a realization is especially vital for people at what is called the decision-making or political level of security planning. Only then can we stake our claim to be a true maritime power.

FUTURE OF AEROSPACE POWER

This nation is truly proud of the achievements of its air warriors in war and in peace, and on the occasion of its Platinum Jubilee, I would like to offer my warmest felicitations to the leadership, past and present of this great Service. Every Indian is reassured by the growing strength and professionalism of the Indian Air Force (IAF); a world-class force with worldwide reach and striking power.

I am faced by a practical dilemma: should I speak as a naval officer or as an aviator? Someone has said: "We are preaching to the converted." I have an advantage here, because naval aviators are reputed to be schizophrenic. They have split personalities because they are neither fish nor fowl. So I will capitalise on my "disability" and speak for both sides.

Let me play the Devil's Advocate and make two points. One is that we are talking about "Aerospace Power" and not "Air Force Power". The army and the navy, therefore too, have a stake in it, even though it may be a relatively small one.

Secondly, history has repeatedly confirmed that air power is decisive in conflict, and there can be little doubt that "aerospace power" will certainly dominate the battlefield. But one cannot overlook the fact that by itself, airpower has not been able to prevail in any conflict. Look at World War II, Vietnam, Gulf Wars I and II, Afghanistan and lately in Lebanon. A major limitation of air power is that it is inherently transient; you can operate "through" the medium of air but not "from" it. Therefore, in a conflict, you cannot do without what the army calls "boots on the ground" and the navy refers to as "forward presence". So let us face it: we all need to move together participatively or jointly.

As Lieutenant General Tebogo Masire of Botswana has said, we cannot look at the world through monochromatic lenses, because each nation and each region is faced by

Adapted from Keynote address delivered at the Valedictory Session of the International Aerospace Power Seminar, February 5, 2007.

its own set of challenges. And just as there is no single panacea for all problems, we need to define a future for aerospace power that is relevant to our conditions and environment.

Aircraft were used in offence for the very first time in 1911, when the Italians bombed the Turks in Tripoli from the air. And World War I was only three weeks old when the first aerial combat took place over France.

This war was to be the first and last major conflict of the 20th century in which infantry made horrendous sacrifices for gaining or losing a few yards of territory; the total casualties of this war, to machine gun fire, poison gas and barbed wire, exceeded those in all previous conflicts put together. Although extensively deployed, aircraft played a generally defensive and not very crucial role in the outcome of war, perhaps due to lack of doctrine and experience.

All too often in history, the conclusion of each conflict has served as a prelude to the next one, and so it was with World War I. The script of World War II was said to have been written in June 1919, in the Treaty of Versailles, and the inter-war years were spent by strategists working out ways to fight future wars with minimum casualties. And of course, air power offered the greatest promise in this area.

At this juncture I would like to review a few events in the history of air power, which can be termed as defining moments because of the momentous impact they had on the course of aerial warfare; and the end of World War I is a good starting point:

- The merger of the Royal Flying Corps and Royal Naval Air Service in 1918 produced the world's first force, independent of army or navy command for the conduct of air operations; the RAF.
- Naval air power became an established reality, with the first aircraft carrier being completed in 1918, thus giving impetus to a new branch of air warfare with profound implications in the years to come.
- Air power captured the imagination of military theorists like Guilio Douhet, Billy Mitchell and Hugh Trenchard, who staunchly advocated strategic bombing of the enemy heartland to shatter his morale, cripple his war fighting ability and thus obtain early victory, with the least casualties.

Between the Great wars there were many wars. The French, Italians and the British experimented with air power to put down insurrections in their colonies. Interestingly, as far back as 1922, the RAF was practicing a concept called "air control" in Iraq to avoid committing ground troops against the local tribes. The Spanish Civil War and the Sino-Japanese wars were the proving grounds for new flying machines as well as new aerial tactics.

All this was a useful prelude to World War II, from which the Marshal of the Indian Air Force Arjan Singh picked out Dunkirk, the Battle of Britain and Normandy landings, as events of significance. I would like to identify three different events which were to have far reaching implications:

- The first one, which heralded the opening campaign of World War II, was the German concept *of Blitzkrieg* or lightning war. Fast moving armoured columns on the ground, supported by furious air assault by the Luftwaffe resulted in the swift conquest of half of Europe in a few months, and set the bar for future army-air cooperation.

- The second was the commencement in 1940, of strategic bombing in the hope of breaking the enemy's will to fight and bringing the war to an early conclusion. While the bomber offensive certainly caused immense damage to life and property, and also to German morale, whether its impact was enough to shorten the war remains a hotly debated issue.

- The third was the two atomic bomb detonations over Hiroshima and Nagasaki in August 1945. Both bombs were delivered by a single B-29, and their destructive power was enough to result in instant capitulation by the Japanese. This heralded a terrifying new capability of air power, and ensured a place for the manned bomber in the armoury of nuclear powers for many decades to come.

The Vietnam War constituted a watershed for deployment of air power in the 20th century and beyond. It saw a decade and a half of slowly escalating conflict, in which huge air assets were committed and the US lost over 8,500 aircraft and helicopters. Smart weapons were tried out for the first time, and tactics evolved for jet fighters, bombers and attack helicopters, which were to transform air warfare.

While the Arab-Israeli wars and the conflicts on our own subcontinent contributed a great deal to the repository of air warfare knowledge and experience, it is the deployments of air power in the Balkans and the Persian Gulf that opened a radically new chapter in air power, as far as stealth, precision weaponry, information dominance and C4ISR are concerned. In fact, the future can be said to have started in the last decade of the 20th century.

The defining moments that I have highlighted were significant in that they contained seminal lessons, both positive and negative, and were instrumental in the formulation of air power doctrines and strategies, and in shaping its future. Most of these lessons were learnt and internalized by practitioners of air power, but humans are imperfect; and there are probably still many mistakes, which get repeated, and many lessons that are re-learnt again and again.

The Kosovo operations, as the RAF Chief Sir Graham Eric Stirrup pointed out and the recent Israeli campaign against the Hezbollah in Lebanon, with the subsequent resignation of the Chief of Defence Staff Lieutenant General Dan Halutz would possibly throw up their own lessons.

Closer home, the growth of our own air force has remained intimately linked with India's evolution as a nation state, and her slow but steady graduation from somewhat Utopian ideologies in the early years after independence, to the harsh world of *realpolitik* today

One of the few Services world wide, to be born as an independent air force, the IAF was cast deliberately by the British in the role of a tactical army support arm. It grew up in the shadow of the Royal Air Force, which retained for itself, the fighters and bombers as well as the strategic responsibilities in this theatre.

The fledgling RIAF was thrown headlong into conflict within ten weeks of our independence, when Pakistan made an attempt to snatch away the state of Jammu and Kashmir by force. Our Spitfires, Tempests, and Dakotas, deployed with imagination, and flown with great skill and daring in difficult weather and terrain, performed a vast array of missions, which resulted in the invader being expelled from most of our territory.

Since 1947, we have seen four major conflicts; and in each case the IAF's crucial contribution to the nation's defence has been marked by visionary leadership, innovative strategy and outstanding gallantly in the air. In its Platinum Jubilee year the IAF can look back with immense pride at its glorious history and inspiring combat record.

Post-independence, while absorbing state-of-the-art equipment from diverse sources, the IAF has ensured that its doctrine kept pace. In consonance with emerging geo-political realities and the nation's security needs, the IAF has been shaped by its leadership so that today it is a strategic force, honed to a fine professional edge.

Air power has been undergoing a steady process of transformation and major drivers have been the quest for low observability (or stealth) and greater precision in weapon delivery. The effectiveness of air power has thus seen manifold enhancement, and today a single F-117 can achieve as much destruction as squadrons of B-17s or F-105s used to earlier.

It is now apparent that the future will bring everything that science fiction writers visualized in books, and George Lucas used to show on screen. Real time cockpit information, directed-energy weapons, space-manoeuvering vehicles, space based radars, and extensive applications of bio and nano-technologies are all just round the comer. Above all, information dominance will set the rules of the game. According to the USAF doctrine: "In the 21st century it will be possible to find, fix, or track and target anything that moves I on the surface of the earth. This emerging reality will change the conduct of warfare and the role of air & space power."

Asymmetric wars involving terrorism, low intensity conflict, and insurrections are going to be far more frequent than conventional wars between nation states. A major challenge for air forces will, therefore, be to adapt some of the advanced capabilities at their disposal to sub-conventional applications, so that they retain their relevance.

General Paul V. Hester, the US Air Force chief has brought up the most interesting concept of multilateral aerospace cooperation. The USN has been proposing a similar concept, which they call the "1000 Ship Navy". But we have to remember that maritime forces have the advantage of inhabiting a medium in which they can live and operate for long durations. As I mentioned earlier, air forces can only transit through the air but will need overseas operating basis to function multilaterally. But, I am sure that ways and means can be found to make multilateral aerospace cooperation viable.

As we look ahead, a couple of legacy issues which have dogged most air forces of the world, and led them to wage a sustained struggle to assert their individuality vis-a-vis the other Services, need to be touched upon.

In what has now become a classical dilemma, the army and the navy presume that air power is just an extension of their artillery, and the air force, should therefore be the handmaiden of land and maritime forces. Surface forces, both on land and at sea are traditionally most concerned about the "immediate threat" in the theatre or battlefield, because they could come under enemy fire in a matter of hours or even minutes. Commanders on land and at sea, therefore tend to develop "tunnel vision" which is restricted to their own, limited areas of responsibility.

The average airman therefore has a conviction, that as long as hostile air elements are prevented from interfering with our surface operations, its tactical deployment should be accorded a lower priority. They feel that the vast potential of air power is best appreciated by an airman and best exploited strategically

We cannot convert people to see things our way by standing on podiums and hectoring them. This is a viewpoint that needs to be understood, discussed, and appreciated by all those in uniform, so that a consensus can be evolved and implemented on the battlefield. I say this because air power is acquiring rapidly escalating capabilities in terms of reach, presence, striking power and information dominance, and will be the key to joint operations.

Another new frontier looming large before us, which could present great opportunities as well as challenges is space. The Air Chief, Air Chief Marshal S.P. Tyagi has of late, been emphasising the need to focus attention on this field, and recommending the formation of an "Aerospace Command". Recent reports of an anti-satellite weapon test by China have served to validate the IAF's concerns.

Today, space capabilities provide us communications, position-fixing, navigation and time, missile-warning, as well as weather and reconnaissance facilities. The future will see growing dependence on space for facilities, which have a crucial impact, as much on military operations as on economic and commercial activities globally. Therefore what happens in space is a pressing issue, which affects all three Services, and must be addressed with urgency.

But we need to tread with caution here, because the subject is beset with complexities. "Aerospace power" can mean one thing if it is just about C4ISR, and something altogether different if it implies weaponisation.

Many mundane and pedantic arguments can be put forth for and against separating air from space, but let us remember that after a few flip-flops between "air power" and "aerospace power", the USAF has finally settled for "air and space power". Of course they have issued a completely separate doctrine, which deals with Counterspace Operations.

I will mention one additional factor. Although militarisation of space is a *fait accompli,* weaponisation of space remains a very vexed issue in international forums. The joint statement by President Putin and Premier Manmohan Singh expressing their strong disapproval is a clear indicator that they do not want weapons in space.

In this context, it was indeed heartening to hear Major General Xinjiang of the PLA air force conveying the reassurance that China too was opposed to the weaponisation of space. If that is indeed so, we need to ponder on the significance of their recent anti-satellite test, and on our own future course of action in this sphere.

The challenge of space must be taken on headlong, but it is necessary that transition from the concept of "air power" to that of "aerospace power" must be preceded or accompanied by sufficient analysis, discussion and debate with the object of evolving a doctrinal rationale and underpinning. As the Singapore Air Force chief BG Ng Chee Khem pointed out, war is not an exclusively military affair, and we must be careful to ensure that the political establishment is acquainted with all the nuances of aerospace power.

It would appropriate to change tack a little here, because while speaking about aerospace power, the term "power" in our minds is usually associated with explosive ordnance and target destruction. But actually it is not always so, because during peace, which fortunately prevails most of the time, aerospace power can apply itself to many benign applications too.

As the IAF has demonstrated over and over again, airmen can deliver emergency relief, humanitarian and medical aid, and food and supplies, to not just one's own countrymen, but also to friends and neighbours. They can also undertake refugee evacuation, disaster relief, search & rescue, and peacekeeping missions with equal ease. Reconnaissance and surveillance too are benign applications of aerospace power in peacetime. It can also make a big contribution to the war against terrorism and LICO. The essence of air power lies in its flexibility and versatility that permit it to switch roles and to access any part of the globe in a few hours.

As I said, we in the Navy too are stakeholders in aerospace power. The IAF has been our friend and mentor, and while building our own air arm, we have been looking at the dramatic growth of IAF capabilities with great delight and anticipation. We know that in the years ahead, our maritime forces are going to operate in the furthest reaches of the Indian Ocean. The Fleet Air Arm is going to be spread thin and our carriers cannot be everywhere all the time. Under these circumstances, we take great comfort from the IAF's trans-oceanic reach and lethal punch, soon to be supplemented by airborne command and control platforms, and perhaps even space capabilities.

In conclusion, I quote Air Commodore Jasjit Singh who has said in a national daily, apropos weaponisation of space: "At the fundamental level, space power has become a pre-condition to control of land, sea and air power." This is a fact; all three Services already rely heavily on air power, and their reliance on space-based assets for operations is going to only increase. The future is aerospace power, and the challenge lies in exploiting it jointly and wisely to maximise national security.

MARITIME SECURITY

A Vision of Maritime India: 2020

"What ship? Where bound?" is the traditional nautical query that goes out on the air when a warship encounters a stranger on the high seas. Recently an Indian Navy (IN) frigate on passage to the Persian Gulf was challenged on radio by a ship of the Coalition Task Force on patrol: "Unknown ship what are you doing in this area?" The IN ship shot back with barely concealed annoyance, "I happen to be an Indian warship in the Indian Ocean. What are YOU doing here?"

This story serves to illustrate two aspects. Firstly, that old mindsets take time to dissipate; guarding the seas East of Suez was historically considered the 'white man's burden', and in some perceptions we are still be seen as interlopers in our own backyard. And secondly, that India has neither acknowledged her role as a pre-eminent maritime entity in the Indian Ocean Region (IOR), nor has it done enough to shoulder the concomitant regional responsibilities.

While others are reluctantly coming to terms with India's quest for her rightful place in the evolving world order, Indians themselves have remained irresolute and diffident in this regard. Consequently, there is a likelihood, that the concept of 'Maritime India' may present a conundrum, to the citizens of a nation who have remained hostage to a continental mindset for centuries.

Therefore, before making out the case for a Maritime India, a decade and a half into the future, it is first necessary to examine the historical context in which a substantive maritime underpinning can be claimed for India.

Our Blissful Ignorance

While we revere our past and have been brought up to blindly believe that we are the inheritors of a great culture and tradition, as a nation we have been sadly remiss in

Indian Defence Review, Jan-Mar 2007, Volume 22 (1), pp. 10-24.

neither researching the past, nor adequately recording our own history. William Dalrymple provides us a poignant reminder of this shortcoming in the Introduction to his celebrated new book, *The Last Mughal.* He refers to the 20,000 virtually unused Persian and Urdu documents relating to Delhi in 1857, known as the Mutiny Papers, that he found on the shelves of the National Archives of India and says, "...the question that became increasingly hard to answer was, why no one had used this wonderful mass of material before."

It was left to conscientious British administrators, Jesuit scholars and other European researchers to unravel our past by learning Sanskrit and studying our sacred Vedic literature. The discovery and collation of India' history and culture by European scholars not only created great respect for it amongst them, but also inspired Indian nationalists to value and cherish their own inheritance, as they had never before viewed India from such a perspective.

However, those who write history also enjoy the license of giving it whatever slant they wish to; and this as we will see, is often at the cost of objectivity.

Maritime History

Almost all works on maritime history from Western sources start with a description of the seafaring tradition of the Mediterranean basin (*circa* 2,500-2,000 BC) and dwell on the sea power of Crete, Phoenicia, Greece, Carthage and Rome. The first reference to the Orient generally relates to the Graeco-Persian war in the fifth and fourth centuries BC where mention is made (with relief) of how Greek sea power thwarted an "attack by Asia on Europe". In this context, Greece was fortunate in having two accomplished historians: Herodotus who is known as the "father of history", and the Athenian historian-soldier Thucydides, who maintained a meticulous record of the Peloponnesian Wars between Athens and Sparta (431-404 BC).

In the fifteenth and sixteenth centuries, Western Europe came once again under threat from the east; this time from sea power of the Ottoman Turks, who brought pressure to bear in the Mediterranean. The Battle of Lepanto in 1571 put an end to this last Asiatic challenge. Historians conclude that this was also the period that seafaring countries of Europe made certain breakthroughs in the fields of ship-construction and navigation, which enabled its sailors to start undertaking long-distance voyages. At the same time advances in gunnery and specialized fighting vessels provided them the means of overwhelming the opposition of 'other races'.

As Paul Kennedy puts it, the Western conquest of the East, was inspired by a mixture of political, religious and economic motives, particularly the latter. Once the Spanish and Portuguese fleets had demonstrated the ease of conquest, and the economic benefits to be gained by such maritime forays, the race was on, with Dutch, French, and English adventurers joining in what became a scramble for trading links, political advantage, proselytisation and loot.

So much for early maritime developments in the West; but it is intriguing to note that nowhere in these historical accounts does one find even a passing mention of India or the seafaring skills of ancient Indians.

Our History

For this we have to turn to the lone voice of Sardar K.M. Panikkar (1895-1963), India's first ambassador to China, who combined in himself the attributes of statesman, diplomat, historian and visionary. Among the large number of his works in Sanskrit, Malayalam and English, is a seminal essay entitled *India and the Indian Ocean*. First published as a monograph in 1945, this treatise is now out of print, and although read and quoted extensively by foreign scholars, it is little known to Indians.

According to Panikkar, for geo-physical and meteorological reasons (currents, prevailing winds, etc.), it was the Indian Ocean, and specifically the lands washed by the Arabian Sea, which saw the first naval and oceanic sailing activity; and European historians err grievously when they assume that the navigational tradition first emerged around the Mediterranean.

Long before seafaring developed in the "limited" Aegean waters, oceanic navigation had become common with the coastal people of peninsular India, states Panikkar. He clinches his argument by stating that: "Millenniums before Columbus sailed the Atlantic and Magellan crossed the Pacific, the Indian Ocean had become a thoroughfare of commercial and cultural traffic between the west coast of India and Nineveh and Babylon (modem Iraq) as well as the Levant (Eastern Mediterranean)."

Panikkar goes on to assert that not only do the earliest Indian literature, the Vedas (1,500 BC), speak frequently of sea voyages, but that much of the materials found in the remains of the Indus Valley Civilization (3,000-2,500 BC), and many products discovered in Mohenjodaro came either from the shores of the Red Sea or the extreme south of India and "could only have been transported by sea." He places Indians firmly alongside the Greeks and the Arabs as ancient seafarers and claims that the Hindus had already in use a magnetic compass *(matsya yantra)* for accurate navigation, and having acquired the skills to build ocean going ships of great strength and durability ventured into the distant reaches of the Arabian Sea.

Though Socotra was perhaps an Indian settlement and Indian communities existed in Alexandria, and in other locations in the Red Sea and the Persian Gulf, maritime activity in the Arabian Sea was confined to commercial purposes only. On the other hand, the Bay of Bengal provided a highway for a succession of kingdoms in the southern and eastern Indian peninsula to embark on cultural, colonization and proselytisation missions to lands beyond the Malacca Straits – as Far East as Japan. Interestingly, Panikkar debunks the commonly held belief that all Hindus had a religious objection to crossing the seas, saying: "it was never true of the people of the South."

Panikkar recounts the continuum of colonization as well as cultural and religious osmosis from India's east coast, by sea to South-East Asia. Starting with the Mauryan emperors, he traces Indian maritime activism through the Andhra, Pallava, Pandava, Chalukya and Chola dynasties. To make his point about intrepid Indian mariners providing continuous cultural sustenance and support from the 'mother country', he refers to the 500-year long dominance of the seas by the Sumatra based Sri Vijaya Empire (of Indian provenance) and to the growth of large Hindu kingdoms and empires in Champa (Siam), Cambodia, Java, and Sumatra from the 5th to the 13th centuries.

From this apogee, India's maritime prowess went into rapid decline, mainly because the Central Asian dynasties, which ruled in Delhi, knew more saddles and stirrups than concepts of sea power.

The arrival of the 20-gun Portuguese frigate *San Gabriele* off Calicut in May 1498 marked the beginning of what Panikkar terms as the 'Vasco da Gama epoch' and commencement of four centuries of 'authority based on control of the seas' by European powers; and not all the daring, valour and patriotic fervour of the Zamorins, Marrakars or Angres could stand up to it.

An Indian Ocean Entity

"Status and symbolism," said George Tanham in his monograph on Indian strategic thought, "matters greatly in Indian society…and Indian Admirals may need no justification or rationale for a powerful navy other than that India's greatness mandates it." Others have alleged hegemonistic intentions on the part of India. With mindsets of this nature, I considered this somewhat lengthy prologue necessary to obtain a correct perspective about our past, and to provide reassurance that Indians are neither interlopers nor parvenus in the Indian Ocean.

We, therefore, need to examine whether in trying to become a pre-eminent Indian Ocean Region (IOR) entity, India seeks merely 'status and symbolism' and hegemony; or is it actually seeking to fulfill a manifest destiny and a tangible need. The British, because they had arrived in India by sea, realized the gravity of the potential maritime threat, especially from their European rivals. Accordingly, they adopted a maritime strategy for India, which was a sub-set of their global game plan to gain and maintain control of all major oceanic choke points worldwide, especially those leading to the IOR.

Our post-independence leadership, for various reasons, developed a utopian outlook, which led to a *moralpolitik* rather than *realpolitik* orientation in our policies and a focus on lofty concepts like 'non-alignment', 'universal disarmament' and 'zones of peace' which were rich in rhetoric but did nothing to further our national interests, gave us a moralizing image and endeared us to no one.

The British have not forgotten that their economic rise and fall had been closely linked with their navy's rise and fall. Shorn of all imperial or colonial trappings, their Defence Doctrine still speaks of maintaining the capabilities required "for independent action to meet inescapable national obligations and safeguard British interests worldwide." Many times the size of this tiny island nation, we need to make a serious assessment of India's own national interests and compulsions in the context of the IOR.

Due to a lack of vision, diffidence, and preoccupation with internal matters we have over the past 60 years embraced insularity and neglected our maritime security. Even if some kind of a 'Monroe Doctrine' was cultural anathema, we should at least have declared our strategic frontiers, and defined a strategy to safeguard our interests within their bounds.

A Maritime Destiny?

India's overarching interests are clearly defined by the need to guarantee a stable and tranquil external environment for two reasons. Firstly, our own people expect that favourable conditions will be maintained for speedy implementation of the nation's lagging developmental process. Secondly, we have an obligation to the international community to ensure that trade and shipping traffic flows unhindered in the IOR sea-lanes.

With ample justification, our strategic frontiers can, therefore, be considered to extend from the Persian Gulf down the east coast of Africa, across to the Malacca Strait and south to the Southern Indian Ocean. In this context, as a peace loving status quo power and a law-abiding nation state with an impeccable record of observing international conventions, we must disregard those who are inclined to cry wolf about 'hegemony'.

Admiral Mahan the renowned American strategist, had specified six conditions, as having a vital bearing on the sea power of a nation: (i) geographical position; (ii) physical conformation; (iii) extent of territory (iv) population (v) national character and (vi) policy and nature of government institutions. Let us examine India against the touchstone of Mahan's conditionalities.

As far as the first three, essentially geographical conditions are concerned, no country — perhaps not even an island state — could be as favourably placed as peninsular India, for the development of maritime power. The next two conditions relate to the commercial enterprise and seagoing proclivities of the populace; and with 11 maritime states and island territories India probably has more seafaring people than the population of most European countries.

It is the sixth and last condition, on which we will need to focus our attention rather sharply.

Having seen that the portents are appropriate and propitious for India to redeem her maritime destiny, we have to recognise that there are a number of distinct strands to the logic, which must underpin her endeavours in the maritime domain. It is necessary that we examine (the threats, and opportunities that constitute) these strands and then weave them all together into a cohesive cord.

States & Non-State Actors

For a threat assessment to be objective, it must recognise Lord Palmerston's dictum about nations having neither permanent friends nor foes, but only permanent interests. An appraisal, divested of sentiment, will therefore show that India and China are going to be competitors for the same strategic space in Asia, and no matter how peaceful their rise or how intense their bilateral trade, a clash of interests cannot be ruled out. It is intriguing, in this context, to note that of her fifteen neighbours, China has painstakingly settled land boundaries with all, but stoutly maintained her claim on Arunachal as well as occupation of Aksai Chin.

The sustained transfer of not just conventional arms but also advanced nuclear weapons technology as well as missiles to Pakistan either directly by China or through the North Korean conduit has no precedent in international relations. In addition, China's strategy of creating a ring of client states right around India has placed us strategically on the back foot. This situation is, however, of our own making, because over the years, China has provided sufficient indications of her plans for 'containment' of India, which we have disregarded.

It should come as no surprise to us if in the next few years PLA Navy ships and nuclear submarines are put regularly into harbours like Chittagong, Sittwe, Hambantota or Gwadar in our immediate neighbourhood. In pursuit of their grand design, the Chinese are planning or in the process of building container terminals in all these ports.

By herself, Pakistan may or may not have been able to do much vis-à-vis India, but as China's surrogate she has received strategic support, and managed to achieve a great deal. And of course the Chinese puppet-masters have manipulated a willing Pakistan brilliantly to checkmate India at minimal cost to themselves.

Developments post 9/11 have garnered for Pakistan, moral and material support from the USA, and further buttressed the position of her military ruler. As a consequence of this implicit and explicit abetment, Pakistan continues undaunted, to be the nursery of religious fundamentalism and fountainhead of nuclear/missile proliferation.

Moving away from our immediate neighbourhood, we also need to factor into our calculus, the substantive presence of extra-regional powers in the IOR. Friendly they may be, but one should never forget that they are in these waters, not for altruistic motives but specifically to safeguard their perceived national interests, economic and strategic.

Should a conflict of interests ever arise, we must be in no doubt that coercive force will be brought to bear on us. Under such circumstances, we have to be prepared to act in our own self-interest. And we must let neither the Hyde Indo-US Nuclear Energy Act, or any similar document, nor the manifold 'strategic partnerships' that we seem to have crafted with other nations ever cloud our vision.

The seven tons of explosives, which created mayhem in Mumbai in 1993, arrived on Indian shores by sea. Today, the Golden Crescent and the Golden Triangle on either side of India are the source of financial sustenance for terrorist organisations like the al Qaeda and Jemmiah Islamiah, which use maritime routes for lucrative narcotic and aims trafficking. The Liberation Tigers of Tamil Eelam (LTTE), apart from its combatant Sea Tiger wing, also has a small merchant fleet, which conducts clandestine trade in South-East Asia to replenish the organization's logistics. Add to this, the freewheeling piratical activity in locations like the Horn of Africa, the Bay of Bengal and the Malacca Straits, and one gets an idea of the vigil that is necessary to maintain order in the waters of the IOR.

Maritime Assets & Liabilities

The Indian Ocean sees about 100,000 ships transiting across its expanse annually. Two-thirds of the world's oil shipments, one-third of its bulk cargo, and half the world's container traffic pass through its waters.

The vibrant economies of China, Japan, and South Korea as well as the rest of Asia-Pacific rely on oil supplies, which emerge from the Strait of Hormuz and transit via the Malacca Strait into that region. Over 70 per cent of our own oil comes by ship from the Persian Gulf. Any disruption in oil traffic could destabilise price levels, resulting in a major upset for the world economy and a setback for our developmental process. As mentioned earlier, India's fortunate geographical location astride Indian Ocean sea-lane gives her a key role in safeguarding their integrity and ensuring unhindered traffic.

India's burgeoning economy, which ranks fourth in the world in PPP, is inextricably linked with seaborne trade. Our exports were about US$ 100 billion in 2005-06 and are slated to double over the next five years. Of our foreign trade, over 75 per cent by value is carried by sea. India's growing merchant fleet is the fifteenth largest globally and operates out of 12 major and 184 minor ports scattered along our 7,500 km long coastline.

Another aspect that presents the prospect of wealth and prosperity and yet contains the seeds of conflict, are undersea resources. The average depth of the Indian Ocean is less than 4 km, and that is the distance, which tantalizingly separates us from a veritable treasure-house of rare minerals, gas and hydrocarbons awaiting exploitation on the ocean bed. India has a mineral rich EEZ extending currently, over 2.2 million sq km (and likely to increase). In many instances, especially in deep basins of the Andaman Sea, technology is the only barrier that currently hinders exploitation of these resources at this moment.

In an effort to diversify resources and ensure stability in supplies, ONGC Videsh has acquired oil concessions abroad in Russia, Myanmar, Iran, North Africa, the Central Asian Republics and South America. These represent investments of several hundred billion dollars in real estate, infrastructure and national resources, which may one day require us to reach out across the seas for their protection.

The frozen wastes of the Antarctic have been attracting expeditions of Indian geologists, meteorologists, oceanographers and others for over two decades now. Our scientific community has established a succession of manned scientific stations, which have yielded valuable data over the years. Their worthy endeavours have been fully supported by the navy, and should this unique continent have anything to yield in terms of mineral or organic wealth (even if 50 years hence) India's stake would need to be protected.

Fish provides 25 per cent of the world's supply of animal protein. The control and management of fishing resources is a problematic area, with most seafaring nations deploying their fleets to more lucrative grounds in the EEZ of other countries. India's EEZ contains an estimated potential yield of 40 million tons of fish. Of this, we harvest less than 10 per cent and the rest are either poached by foreign trawlers (especially in our island territories), or die of old age.

Maritime Cooperation

During peace, which fortunately prevails most of the time, the main business of navies is (apart from preparing for war) to act as instruments of state policy in offering "a range of flexible and well calibrated signals" in support of diplomatic initiatives. The options available include, projecting maritime power for intervention, or influencing events on land, showing presence to either convey reassurance or threat, cooperating with allies in training exercises or simply rendering humanitarian relief when required.

Recent experience, including the relief rendered by us during the 2004 tsunami disaster and the Lebanon crisis in 2006 has shown that our neighbours are inclined to look instinctively to us for assistance in times of distress. Even in the normal course, they feel that India is well placed to provide military training as well as material aid to them. It is only when we fail to respond to their repeated appeals that they turn to other countries in the region. Regrettably, this scenario continues to be repeated with depressing regularity.

We have been neglectful of this aspect and need to make early amends. Our foreign cooperation objectives should essentially aim to curb or prevent powers inimical to India from intruding into our neighbourhood, and to help us shape the environment favourably for operations in peace and in war. The Indian Navy does have in place a well-oiled mechanism as well as long-term plans for foreign cooperation, which needs the Government's backing for implementation.

Safety of the Undersea Deterrent

India's Nuclear Doctrine clearly envisages a deterrent in the form of a triad with land-based, aircraft borne, and submarine launched 'legs'. Of these, we possess only the first two at the moment. Nuclear weapons are not meant for war fighting, and achieve deterrence by convincing the enemy of the futility of contemplating a first strike, because the instant response would be so horrific and devastating as to render the strike pointless.

In order to convince the enemy, your deterrent must have two essential attributes, which render it 'credible.' It should have massive destructive power, and a major component of it must be survivable in the face of a pre-emptive first strike. The only platform, which can claim to be 'undetectable' and hence invulnerable to pre-emptive attack, is the nuclear propelled, ballistic missile-armed submarine (known in USN parlance as the SSBN), which can remain concealed in the ocean depths for months on end.

The waters of the Indian Ocean provide a sanctuary for the SSBNs of various nuclear powers, including the Chinese. These vessels lurk unseen with their missile warheads programmed to strike designated targets, some of them, no doubt, our own cities. Location and tracking of these submarines may become necessary to keep the threshold of coercion at a reasonably high level. This is however, a daunting task, which requires tremendous anti-submarine warfare hardware and skills, which the IN should be acquiring.

To complete the triad of our own Strategic Forces we must have a small number of Indian SSBNs. Developmental work is reported to be underway, and when this platform becomes operationally available, we will need suitable areas in the distant reaches of the Indian Ocean from where it can be safely deployed to pose deterrence to our adversaries.

The Ingredients of a Maritime India

At Independence, agriculture generated 70 per cent of India's GDP, and it is a sign of the times that today its share has dwindled to less than 20 per cent while services (including IT) contribute over 50 per cent; and industry does the rest. The dramatic growth of the Indian economy is being spurred by its interdependence on, and integration with the global economy. This factor, coupled with our energy requirements, burgeoning trade, oceanic wealth; both mineral and organic, and many other vital interests require us to focus attention on our maritime environment.

A supporter of all UN organs, and an aspiring member of the Security Council, India's own well being and progress depends on promoting international stability, freedom and economic development. Since our economy is dependant on international trade, India's vital interests are not going to be confined merely to the IOR. Just as we see foreign direct investment pouring into India, Indian investment overseas is also going to grow rapidly. Thus, along with an Indian diaspora of over 20 million, we are also going to have vital economic interests scattered worldwide.

The concept of maritime power encompasses far more than most people seem to imagine, and certainly goes much beyond the military aspects. Although it may be no longer fashionable to quote Admiral Sergei Gorshkov, in the opening pages of his book *Sea Power of the State* he highlights his expectations from maritime power as follows: "In the definition of sea power we include as the main components, ocean research and exploitation, the status of the merchant and fishing fleets, and their ability to meet the needs of the state, and also the presence of a navy to safeguard the interests of the state since antagonistic social systems exist in the world. Sea power emerges as one of the important factors for strengthening the economy, accelerating technical development and consolidating economic, political and cultural links with friendly people and countries."

Thus, contrary to popular perception, a strong and capable navy is just one (albeit very important) component of a nation's maritime strength. We need to focus on the ingredients required to make us shun our centuries old continental mindset and put us on the path of becoming a truly maritime nation. Let us turn our attention to the different ingredients that will go towards constituting a vibrant Maritime India in 2020.

Ports and Merchant Fleet

Considering that 97 per cent of our international trade by volume, is carried by sea, the maritime rector, in which the Ministry of Shipping and Transport includes port operations, the mercantile fleet and our shipbuilding industry, has been sadly neglected since Independence. A study commissioned by the Confederation of Indian Industry (CII) in 2006, to examine the revival of this sector, points out that with our seaborne trade rising at a rapid rate, there is urgent need to focus *inter alia* on the following areas:

- Ports
- Global Maritime Security Environment
- Hinterland Connectivity
- Shipbuilding and Ship Repair Industry
- Human Resource Development

Compared to the efficient cargo handling and speedy ship turnaround times available in most of Asia-Pacific, our ports are slothful and grossly inadequate to meet the current cargo throughput requirements. Considerable planning and investment would be needed to bring our ports up to international standards. In this context, the exacting requirements of security protocols like the ISPS Code would also need to be kept in mind. Moreover, unless hinterland connectivity in terms of efficient railroad and fast highway connections are available, investment in ports may be rendered infructuous.

7India's merchant fleet comprises of 760 ships of 8.6 million tones GRT, and the average age of its vessels is about 17 years. It can carry less than half of the country's foreign trade, and India's shipping capacity in the words of Rahul Roy-Chaudhury is "...woefully inadequate, and by any reckoning out of synch with its overwhelming dependence on seaborne trade." Consequently we are continuously dependant on foreign carriers and losing earnings to them. The fleet is also qualitatively mismatched to market needs; lacking container vessels, as well as product and specialized carriers. All these shortcomings constitute strategic handicaps and need to be redressed.

Shipbuilding

Of all the Indian flagged vessels, only about 10 per cent have been built in Indian shipyards because of higher costs, lengthy delivery times and indifferent quality. There is deep irony in this statistic, because we have an ancient shipbuilding heritage. In Lothal (Gujarat), archeologists have excavated, possibly the world's oldest dry-dock going back to 2,500 BC, and anchored in Hartepool Harbour, UK is the 38-gun frigate HMS *Trincomalee* built of stout Malabar teak by the Wadia master-shipbuilders of Bombay way back in 1817. Today, India's shipbuilding industry is a pale shadow of the magnificent Wadia tradition.

India is at the centre of a spectacular IT boom and many industrial sectors in the country are at the cutting edge of technology or of production engineering, but not shipbuilding. It is an index of our shortsightedness that while China, South Korea and Japan have marshaled their strengths to produce quality ships at competitive prices in large numbers, India with all her advantages, has completely neglected the shipbuilding industry. India may be world number one in low-tech ship breaking, but a modern supertanker needing repairs in our waters may well have to go to Dubai for dry-docking.

Of the 28 established shipyards in the country, only seven public sector and two private yards have reasonable building capacity. While the public sector shipyards lack the technology, as well as finances, work ethic and innovative spirit necessary to be competitive, the private shipyards await a 'level playing field' to make their mark. This depressing scenario may persist unless the Government of India takes a long-term view and implements some hard decisions in this strategic sector, bearing in mind that a boom in shipbuilding will spawn a multitude of ancillary industries too.

A Central Maritime Agency

The sixth issue figuring in Mahan's agenda referred to earlier, spoke of 'government institutions'. Today there is no single government agency, which has either the span of responsibility or the authority to act as the focal point for India's maritime policies and interests. Nor one, which has the physical means to exercise control over the myriad activities that take place on and under the oceans. As many as sixteen different ministries, departments or organisations, (including the Indian Navy and the Coast Guard), are involved in ocean-related matters, and much of the time the left hand does not know what the right is doing. The result is; confusion, crossed wires and compromised national security.

Let us take a few examples. Merchant ships blatantly pollute our waters with impunity, unsafe ships ply in our waters and often run aground or sink outside harbours, licenses to fish in Indian waters are given for foreign trawlers who fudge papers and have Chinese or Pakistani crews, we lack radar chains to monitor shipping traffic, and the crowning irony; while hundreds of retired Indian naval personnel are debarred from crewing our merchant ships, we look abroad to hire foreigners for this purpose.

A comprehensive proposal for the constitution of a multi-disciplinary 'Maritime Commission' was mooted a few years ago by Naval Headquarters, but ran into rough weather and finally foundered on the rocks of inter-Ministerial insecurity and rivalry. A nation such as ours urgently needs to evolve an overarching Maritime Policy and create a central agency to monitor its implementation.

Varuna's Trident

Coming finally to the Indian Navy, which is the instrument for safeguarding our maritime security perimeter, for creating a position of influence in the region where India's national interests lie, and *in extremis* for defeating the nation's enemies at sea. Of all the entities that have found mention so far, it is the IN, which has prepared itself best to be the keystone of the nation's maritime edifice in 2020.

Therefore, it is necessary to discuss the Navy's outline plan of action for this period. The Navy considers that the resurgence of our maritime power is a *sine qua non* of India's rise as an economic giant. While undertaking the planning process, the Navy's leadership has taken care to create not just the intellectual underpinning necessary for it, but also to provide, for decision-makers, a rationale for its projected growth path.

The *Indian Maritime Doctrine* published in 2004, essentially set out the 'rules of the game' for deployment of maritime assets for attaining national objectives. In 2005, the navy decided to undertake an exercise to quantify the 'capabilities' (for example air defence, amphibious, anti-submarine, maritime patrol, etc.) that would be needed to discharge all the roles envisaged for maritime forces in the doctrine. Given the performance of modern ships, aircraft and submarines, these capability requirements were then translated into numbers of platforms that would be necessary.

A sensitivity analysis was undertaken with the numbers that emerged, against budgetary variables up to the end of the 13th Plan (2022). Having essentially met all bottom lines relating to threat perceptions, fiscal resources, and shipbuilding capacity, this force planning exercise was converted into a *Maritime Capabilities Perspective Plan* which now forms the Navy's force planning blueprint till the end of the next decade. This capability-based approach has served to make tomorrow's navy leaner, while packing far more punch and keeping the 'capital to revenue' ratio at a very healthy level; which means that there is much more money available for modernization.

In late 2006, the IN promulgated a Maritime Strategy. This document has served to fill the residual philosophical hiatus, and to provide tangible guidelines, within the specific geo-strategic environment anticipated in the next decade, for the acquisition, build up and employment of maritime assets in peace and in war.

The navy of 2020 will essentially be a three-dimensional force ("Varuna's Trident") built around the core of two aircraft carrier task forces and closely networked through a dedicated communications satellite. Indigenously built destroyers and frigates will be available in adequate numbers to provide escorts for the carriers as well as for independent surface action and anti-submarine hunter-killer groups. All escorts will have modern sensors and long-range weapons of offence and defence, and will carry multi-role helicopters.

While replenishment ships will ensure long legs for the combatants, we will also have enough friendly refueling ports in the IOR (and South China Sea) to allow extended operational reach. Long-range anti-submarine warfare and maritime patrol aircraft of a new generation would provide support to our forces in the distant reaches of the oceans.

The Navy would have in service by then, six indigenously built submarines of the Scorpene class, and perhaps another 3-4 boats of an advanced indigenous design, all equipped with missiles and air independent propulsion. The decline in submarine force levels should have been arrested and reversed.

It is now an article of faith with the IN that all operations by maritime forces at sea will be designed to produce a direct or indirect impact on the land battle. Future operations will invariably demand that information dominance be the opening gambit. Sea control, if required, could then be established as a prelude to maritime manoeuver and littoral warfare. In such a scenario, land attack, naval air, amphibious, and Special Forces capabilities will require due emphasis, which is being provided by the naval planners.

Many of these concepts are new, and require radical reorientation of mindsets as well as operating procedures. The Navy has, therefore, triggered a process of 'Transformation' to deal with orderly the management of change.

The Navy's focused thrust would have ensured by 2020 that much technology and many products of indigenous or collaborative origin are at sea. In the area of propulsion, we would have advanced diesels and gas turbines, as well as electric drives on our warships. It is entirely possible that the endeavours of our scientists would by this time have succeeded in putting indigenous nuclear propulsion at sea, either on a submarine or an aircraft carrier.

Conclusion

It may be mentioned with utmost emphasis, that of all the ingredients, which go into the making of a great maritime nation, none is more important or significant than the human mind. Unless determined efforts are made to create a consciousness of our ancient maritime heritage, and an affinity for the seas in the minds of young Indians, all efforts at creating a Maritime India will come to naught.

There will be skeptics who point to our populous landlocked states. But remember that while the Indian Navy today does have its share of sailors from the 'maritime states', the bulk of our Service consists of land bound people from Bihar, UP, Rajasthan (two former Chiefs), Haryana, Himachal Pradesh and Punjab; and they all make excellent sailors. So much for Mahan's predictions!

In an epic and hazardous voyage, the Navy's sail training ship INS *Tarangini* circumnavigated the globe in 2004. The Captain describes with delight, the incredulity of thousands of visitors who boarded the ship in various ports across the world at the sight of Indians sailing around the world. By next year, we hope that a courageous Indian Navy officer will set another seafaring benchmark for his countrymen by undertaking a solo sailing voyage around the world.

There is also the example of continental Russia where Peter the Great almost single-handedly created not just a maritime tradition but also a magnificent navy, which celebrated its 300th anniversary in 1996. So if a maritime tradition can be created, it can certainly be revived.

Maritime Dimensions of A New World Order

At the outset, I must mention the names of two old friends present. The first one is that of Admiral Takashi Saito, the head of the Japan Maritime Self Defense Force (JMSDF). He and I were in the US Naval War College of 1990. The other person is Professor John Hattendorf, Professor of Maritime History and my guru in the same college. From him I received not only sage advice and wisdom, but also much needed encouragement. To both of them I owe a special thanks for their presence today.

I think the key to our present predicament may well lie in how we interpret the phrase 'new world order', which has tended to surface every few decades since the 1920s with a different connotation each time. It was first heard after World War I when it referred to the rosy but short-lived prospect presented by the League of Nations. After World War II, it came briefly into vogue again in the context of the UN and its organs, and occasionally included the Marshall Plan and North Atlantic Treaty Organization (NATO) as its manifestations.

In the post-Cold War period, Mikhail Gorbachev is credited with reviving the term in an address to the UN in 1988 in an idealistic and wide-ranging vision based on the integration of the Soviet system into the international order. It has thereafter been redefined frequently; mostly by those who feel that they have the wherewithal and strength to change the existing world order. The two Gulf Wars and certainly 9/11 have added new dimensions to these formulations.

I have no intention of even attempting to define the 'new world order' because, that, is best done by the experts and scholars. Unless entropy prevails, our future environment is certainly going to be defined by the world order that emerges. So I would like to flag a few factors, which would have an influence on this issue, and perhaps an impact on the maritime dimension.

Adapted from a speech delivered at the inauguration of the first international seminar organized by the National Maritime Foundation, New Delhi, February 16, 2006.

A decade and a half ago, dramatic events were unfolding before us and were shaping the world. I happened to be studying issues of national security in the Naval War College in Newport. In early 1989, Soviet forces withdrew from Afghanistan, and in November that year, the Berlin wall fell. In February 1990, the Soviet Communist Party agreed to give up power, and the Commonwealth of Independent States was just months away.

Writing in the summer of 1989, when Communism was everywhere in retreat, Francis Fukuyama, a young Deputy Director in the US State Department, wrote in an article: "What we may be witnessing is not just the end of the Cold War, or the passing of a particular period of postwar history, but the end of history as such: that is, the end point of mankind's ideological evolution and the universalisation of Western liberal democracy as the final form of human government."

Seventeen years down the line, the world is indeed a different place, and Fukuyama an older and perhaps wiser man. History has certainly not ended, nor is Western liberal democracy anywhere near being the universal norm. In fact, we in India, as a steadfast democracy, tend to feel a bit lonesome when we look at the political horizon around us. In this context the emergence of Hamas in Palestine, the Hizbollah in Lebanon, Muslim Brotherhood in Egypt and a conservative regime in Iran are significant trends that needs to be noted by Fukuyama.

In 1993, Samuel Huntington posited that the world was entering a new phase, in which the great divisions among humankind and the dominating source of international conflict would not be ideological or even economic, but cultural. The fault lines of civilization would be the battle lines of the future, he said, and in the emerging era of cultural conflict, the US must forge alliances with similar cultures and spread its values wherever possible.

This thesis was received with a degree of scepticism, but now many people are holding up issues such as the civil unrest in France, the Sydney riots, and the Danish cartoons to indicate civilizational cleavages and to highlight the inevitability of conflict.

In India, despite ethnic, linguistic and religious diversity we are proud of our composite cultural heritage, which is the powerful glue that binds us as a nation. While Indians in general have firmly rejected Huntington's thesis, we have to remember that India is also located at the junction of three civilizations, Islamic, Hindu and Sinic, and must therefore tread with caution if we are to safely negotiate these so-called "fault lines".

Let us look at some economic factors and their likely impact at sea. Globalization of the world's economies has brought about so much interdependence that investments worth billions are being made by nations cutting across political and ideological divides. Political dogma has taken a backseat, in the process. In 2004, merchant ships carried ninety per cent of world trade, worth US$ 8.3 trillion. Of this, India's share presently hovers at just below one per cent. With our GDP rising at over 8 per cent annually and trade contributing 23 per cent of it, our seaborne traffic is set to rise sharply, and with it our dependence on the seas.

Yet, despite the success of globalization, a new protectionist movement is gaining strength. In a strange reversal of roles, emerging economies, once vocal opponents of open markets are emerging as champions of globalization while many in the West are becoming panicky because now they are faced by low priced competition and outsourcing of jobs. Economics has historically been an area of potential tensions, and could become the root of future conflict.

Rapid growth in economies has led to a disproportionate rise in energy consumption, particularly in the energy-deficient industries of China and India. This has triggered off a race to secure new oilfields through overseas investment and exploration. India has acquired gas and oil fields extending from Sakhalin in the Far East to Sudan, and even Brazil in distant South America. China and India are pitted as rivals in the market for the best deal in oil or gas assets, and despite attempts at cooperation, the highest bidder will always win.

With hydrocarbon interests becoming transnational and extending worldwide, pipelines running across national boundaries and under the sea, it is anyone's guess what kind of tensions and conflicts of national interest the future will bring. But one thing looks certain: navies will play a key role in energy security. So, even if we disregard Huntington's thesis, I would say that the stakes are high, and seeds of conflict do lie embedded in the IOR. We would need to navigate with care if peace and tranquility is to be maintained.

Where does India stand in all this? To obtain a correct perspective of our position, there is need to stand back and get a historical perspective.

For many years, Indians have deluded themselves into believing that India is a continental power, and we as a nation have proceeded on that assumption. The fact is that from the 3rd century BC till the 13th century AD, India was a thriving and dynamic maritime entity.

While ports on the west coast undertook intensive commercial activity with the Persian Gulf, Red Sea and the Mediterranean, successive kingdoms in peninsular and eastern India displayed a powerful maritime tradition and vision. Dynasties like the Mauryas, Satavahanas, Pallavas, and Cholas sent out fleets that were instrumental in spreading India's trade, culture, and religions by sea to South East Asia and further. The Sri Vijaya dynasty set up a surrogate kingdom in Sumatra.

Consequently, India's cultural bequest is evident in the language, dress, customs and food of Myanmar, the Malaysian peninsula, the Indonesian islands of Java, Sumatra and Bali, as well as Laos, Cambodia and Thailand. It was the decline of our maritime power and tradition in the 13th century AD that coincided with the domination by foreigners for the next 600-700 years.

It is therefore not only fortuitous but, most appropriate that India's maritime resurgence should coincide with her rise as a significant economic power. People speak of India's "manifest destiny" or a regional sphere of influence like the "Monroe Doctrine", but such concepts do not actually figure in our calculus.

We have learnt through bitter experience that it is not possible to undertake economic development or the social uplift of our masses, unless we can ensure a secure and stable environment for ourselves. Moreover, our geography, economy, population and a host of other factors including our dependence on the sea and our vast maritime resources predicate our position in the IOR. We are quite clear that the Navy's two prime roles are to safeguard the nation's maritime interests, and to ensure peace, tranquility and stability in our region.

Neither mathematical models, nor computer simulations; not even crystal balls can accurately predict what the 'new world order' is actually going to be like, because there are so many imponderables. I do not think we need to go any further than Iraq to prove this. However, having swept the maritime horizon through our 'search periscope', and made an assessment, we in the Indian Navy have taken some basic decisions.

Firstly, we will shape our navy, taking into account, the polarization of forces and the current and future maritime capabilities in our region. And here we will look at the full spectrum of conflict, from low intensity warfare to strategic deterrence.

Secondly, we will maintain a capable maritime force, which should have the reach, endurance and sustenance to safeguard our maritime interests in areas the IOR. It will essentially be formed around the core of two aircraft carriers, with necessary escorts, and adequate submarine and aviation components, and a logistics train.

Thirdly, we will do our utmost to foster self-reliance, and this maritime force will be created through indigenous design and industrial effort.

And lastly, we are determined that in consonance with India's ancient maritime tradition, this will be a force for peace, friendship and goodwill, which will reach out to extend a helping hand, wherever needed in our maritime neighbourhood. As the tragic tsunami of 2004 showed, in the midst of our own preoccupations, we had the capability, and the resolve to come to the aid of our neighbours. This will be our leitmotif for the future.

LIGHTING A LAMP FOR MARITIME ISSUES

A lack of strategic vision has been a cultural shortcoming, which has not only had an impact on India's destiny, but also shaped the nation's history. George Tanham, an American analyst who undertook a study of Indian strategic thought, came to the conclusion that there were three historical reasons for this unique and glaring lacuna:

- Firstly, because India had lacked political unity through most of its history, we had not thought in terms of national defence planning.

- Secondly, the Indian concept of time, or rather the lack of a sense of time -Hindus view life as eternal — discourages planning.

- Thirdly, Indians consider life a mystery, largely unknowable and not entirely under man's control. In Indian eyes, therefore, man can neither forecast nor plan ahead with any confidence.

Most of us would scoff at these statements as ridiculous. However, it is an outsider's view of us, and may partially explain why we are, where we are today. Whether we accept George Tanham's analysis or not, one of the more tangible symptoms of this shortcoming has been an affliction that we in the Navy call "sea blindness". This is what has led us to ignore our history and the past, and to push to the margins, most issues relating to the maritime dimension of the country; especially our maritime security.

Indicative of this is the fact that today, maritime or sea-related issues, all of which impinge on our security, can fall within the purview of one or more of 40 odd Departments of the Government of India which deal with such matters. It is for this reason that decisions on important maritime issues, are either interminably delayed or go by default.

Adapted from a speech delivered at the inauguration of the National Maritime Foundation, New Delhi, February 15, 2005. The Defence Minister, Pranab Mukherjee inaugurated the NMF.

And that is why the Navy has been trying to pursue the constitution of a National Maritime Commission, which continues to hang fire.

India is in every respect, a maritime nation; we meet most of Mahan's criteria, and we are the only country to have an ocean named after it. However, there has been a degree of reticence and inaction regarding crucial maritime issues – arising largely from ignorance of such matters. A time has now come, when we ignore the sea only at our peril.

To take just one indicator; Kamal Nath, the Commerce Minister, has informed the Parliamentary Consultative Committee, that India's exports for 2004-05 are likely to cross US $75 billion, and would double in the next five years. Over 90 per cent of this trade is seaborne, and thus there is a close link between the sea and our economic well being. One need not even mention our dependence on tanker borne oil traffic

The need for creating a think tank as a lamp to shed light on maritime issues has been felt acutely by all of us for many years now. But all sorts of problems - some real, and some imagined - kept cropping up, and for many years we sidestepped this issue. The main impediment was that such institutions must essentially be autonomous and non-governmental in nature, and the Navy was therefore constrained in its scope of action.

It was an initiative by a set of public-minded members of the Navy Foundation last year, which triggered the present chain of events. The deal was quite straightforward. If the Navy Foundation donated their time and intellectual abilities, the Navy would look after the more mundane and material matters.

While we were mulling over this proposition, an agonizing question kept cropping up before us. Should we get around to creating such an institution, who would we find to head it? The solution came to us suddenly and almost miraculously. One evening, at a dinner in this very building (Naval Officers Mess Varuna), Vice Admiral K.K. Nayyar ambled across to me, plate in hand, and said in his normal dramatic fashion, "If you create a maritime foundation, I promise to devote five years exclusively to it." On words to this effect, I grabbed the offer before he could change his mind.

This was actually a defining moment, which galvanized all of us into urgent action. I would like to convey the Navy's profound thanks to Admiral Nayyar for making himself available and lending his distinguished name for this excellent cause.

It has always been my belief that the dividing line between the serving and the retired fraternities of the navy is very thin and blurred. And that is how it should remain, because retired officers, have the potential to make an invaluable contribution to the Service in their own way. They are free of many of the fetters that we wear along with our uniform, and therein lies their strength. I have also believed that the Navy Foundation had much more to contribute to the Navy's cause than just inward looking annual meetings, or the odd social gathering. The NMF should provide enough scope for meaningful contribution by the Navy Foundation.

It is our earnest expectation that the National Maritime Foundation will grow into an independent and open forum for exchanging ideas, and disseminating knowledge on maritime issues amongst our countrymen. We hope that it will also become a repository of knowledge and expertise in maritime affairs, and through the medium of a fine journal, and periodic seminars and workshops, stimulate original thinking on these matters.

India's Quest for Maritime Security
Flagging the Route

Crystal gazing and trying to envision the future has of late become an essential and rewarding occupation. Not so much in our country, but elsewhere in the world, large amounts of financial and human resources are dedicated to undertaking net assessments and "scenario building" exercises so that the future does not bring any nasty surprises. While information in this age is plentiful, and there are many tools available for forecasting the future, prognostication remains a rather hazardous undertaking due to many imponderables, of which technology is just one.

With all her think tanks and oracles of wisdom, would the USA have embarked on the current undertaking in Iraq if she had known the quagmire it would become? For all the disaster simulations that must have been done, could anyone have ever imagined that a hurricane hitting America's deep south could expose such shameful organisational flaws and racial fissures in that powerful country?

Consequently, while it is useful to study trends and extrapolate them into the future, it is just as important to concentrate on certain fundamentals. Whether it is international relations, economics or security issues, we need to keep the basics in mind so that we do not "miss the woods for the trees" or "the trees for the wood" and come to misleading conclusions. In the military arena planning is guided by what we call the 'Principles of War'. They constitute the distilled wisdom from centuries of warfare, the observance of which is likely to bring victory in battle.

In a similar manner, there exist some fundamental axioms, as far as international relations, economics and national security are concerned, which help nations to steer a safe and steady course, notwithstanding the uncertainty and entropy that prevails in such affairs.

Adapted from an address delivered at Calicut University while inaugurating a seminar on "India's Maritime Perspectives – 2020", at Calicut, September 5, 2005.

The first axiom, which was voiced by Lord Palmerston states that, "Nations have neither permanent friends nor permanent enemies, only permanent interests" and the sole criterion for a country's security and foreign policies must therefore be self-interest. Unfortunately, India has too often ignored this dictum, and allowed idealism or optimism to cloud our vision. Having shunned realpolitik for what we perceived as a moral or principled stand, we have frequently found ourselves isolated and friendless. In all our relationships, including those with the major powers like USA, Russia or China, we have at some time or the other, faced disillusionment because our expectations were based on idealism, whereas their reactions have been based on realism.

Similarly, while attempting to tackle Pakistan's ruthless proxy war, or their blatant nuclear proliferation we have continuously sought approval of our stand, and condemnation of Pakistan by others. To our disappointment, this has frequently not happened because others view events through the lens of their own self-interest. I dare say, an analysis of our relationship with Nepal, Bangladesh and Myanmar will show up similar flaws. The point I seek to make is that our stand and our actions must be guided by pragmatism and our own supreme self-interest, even if it means that idealism takes a temporary back seat.

Stalin, when informed of the Pope's disapproval of his policies, is famously supposed to have asked, "How many divisions does the Pope have?" The second axiom can therefore be colloquially stated as, "Speak softly, but carry a big stick". This relates to backing up of a nation's intent with the necessary strength. Neither economic prowess nor military might by themselves can count for much in international affairs. It is a combination and synergy of both that will make an impact. I might mention that the concept of strength actually has two components: one relates to military capabilities, but the more important part is the national will or resolve to act in a crisis.

In a related context, it is illuminating to recall one of the most daring but blatantly aggressive acts undertaken in peacetime by one sovereign nation against another: the 1981 Israeli raid on Iraq's Osirak reactor. It was breathtaking in its audacity and violative of all norms of international conduct but the Israelis got away because of another unwritten axiom, which says: "Nothing succeeds like success."

Finally, there is the need for us overcome our preoccupation with internal matters. I think that it is only a matter of time before we set our house in order and prevail over the numerous internal security challenges that confront us today. Therefore, the time to start looking outwards and broadening our horizon in the context of international relations is now. When we look beyond our shores, we will need forces, which have transnational capabilities, and contingency plans to match. The least offensive and optimal instrument of transnational capability is maritime power, and the ideal medium to convey it, are the oceans.

And do remember that there is nothing aggressive or threatening about such a capability; for if we did not have it in a small measure, we would not have been able to render aid in the tsunami crisis. I might mention that at the time of Hurricane Katrina an IAF IL-76 was standing by at Palam, with medical personnel and rescue teams to fly across to New Orleans, should the US Government have accepted our offer.

The stakes for this country are enormous. Today, India is the world's fourth largest economy in terms of purchasing power parity. In the latest study on India by the Carnegie Endowment, titled appropriately enough: *India: A Global Power*, we are described as one of the four 'power centres' in the world. Our vast and vibrant population, which used to be considered a handicap, only a few years ago, has emerged as a tremendous advantage. Our geographic location, particularly in the maritime context, puts us at centre-stage in today's strategic scenario dominated by needs of energy security and terrorism concerns. India's emergence as a global power has thus moved from the realm of 'if and 'maybe' to the question of 'when'.

The maritime domain is playing a significant role in our renaissance, and there should be no doubt in anyone's mind today that India needs to pay particular attention to matters relating to our maritime interests: be it the EEZ, the merchant marine, the fishing industry offshore oil and gas production or deep sea mining. For the foreseeable future, the bulk of our international trade and our energy supplies will remain dependent on vital sea lines of communication. We will also turn more and more towards the seas as a repository of resources and food.

Maritime power, principally in the form of the Indian Navy, supported by the Indian Coast Guard, will play a vital role in defending our maritime interests. The Navy will also play an important role in supplementing, and in some cases, spearheading our diplomatic efforts, particularly in the Indian Ocean Region. Interest in the Proliferation Security Initiative has revived, and it is quite possible that we may be shortly invited to join. Maritime terrorism, gunrunning, drug smuggling and piracy are all current threats, which will need to be countered by determined and coordinated international maritime action.

MODERN NAVIES
A Force for Good

As the end of the Cold War hove into sight, a bright young civil servant named Francis Fukuyama declared from his "crow's nest" in the US State Department that this defining event presaged, "...the end of history as such..." and foresaw a utopian vision of universal liberal democracy. In response to criticism, Fukuyama was quick to clarify that he did not imply a future world free of tumult, political contention, or social problems. But even he could not have foreseen either the dogged persistence of authoritarian forms of governance, or the kind of tumult that would be wrought by non-state actors on the geo-political arena. Far from being the "epilogue", this was to be the beginning of a turbulent new chapter in world history.

So what kinds of roles do modern navies, which are the core of today's maritime forces, have to play in the chapter of history, which now unfolds?

In the sharply polarized Cold War era, navies collaborated largely within their political groupings, like NATO and the Warsaw Pact, while those "outside the pale" like the Indian Navy (IN) had no choice, but to go it alone and remained insular. The walls having crumbled, the past decade and a half, has increasingly seen Mahan's vision of the seas becoming a "...great highway; or better, perhaps...a wide common" come true.

Globalization of the world economy and the resulting trade and energy interdependence brought home the realization that navies needed to synergize their efforts to ensure tranquility on the high seas for universal benefit. This trend has gained particular momentum after 9/11, with the focus now additionally on combating international terrorism and proliferation of WMDs; these twin menaces finding the "great highway" a most convenient medium.

In peacetime, the policy of the IN envisages active constructive engagement with both regional and extra-regional navies, not just for mutual professional benefit but equally to build bridges, and to establish interoperability. Specifically, this takes the form of regular Joint Exercises with the US, Russian, British, French, Singaporean and Omani navies and Maritime Assistance to regional navies in the form of training, consultancy, hydrographic surveys, material assistance, EEZ patrol, offshore security etc.

The IN also recognizes the need for multilateral efforts to combat maritime manifestations of international terrorism, arms smuggling, drug trafficking, piracy and illegal immigration. As a result, we have initiated maritime cooperation with countries of the Indian Ocean littoral and carry out Joint Patrols with the navies of Sri Lanka and Indonesia. We have on request, escorted high value USN ships and submarines in the Straits of Malacca during 2002, and provided warships for seaward security of Maputo, on two occasions, in 2004, for the Government of Mozambique.

In their "benign" role, IN ships have been regularly deployed at short notice, both in India and abroad, to render assistance for disaster relief, search and rescue, diving and salvage, hydrography, and even historical and marine scientific research. The smooth conduct of massive post-tsunami relief operations have served to validate the cooperative endeavours of the Indian Navy in the IOR over the years. Having rushed aid to our stricken east coast and the Andaman and Nicobar island group, we were able to turn simultaneously to our neighbours in their hour of need.

Within a few hours of the disaster, the first naval aircraft with medical teams and medicines landed at Colombo, and by the end of the day the IN had deployed 19 ships, four aircraft and 14 helicopters in the tsunami affected areas. In this, the biggest ever humanitarian relief operation undertaken by the IN, a total of 38 ships, 21 helicopters, 8 fixed wing aircraft and about 5,500 personnel were eventually mobilized and deployed afloat and ashore, in Sri Lanka, Maldives and Aceh in Indonesia.

In the years ahead, the IN will move "full ahead" to establish mutually beneficial collaboration with our maritime neighbours. While the time to "beat our swords into ploughshares" may not have yet come, in my view, a Navy capable of fulfilling its "Military Role" will automatically be capable of executing its less arduous peacetime roles. However, establishing naval cooperation often means the need to prepare the ground and to go that extra mile - and the Indian Navy intends to do just that.

FUTURE STRATEGY AND CHALLENGES FOR THE INDIAN NAVY

During my interaction with other navies, I often discerned a degree of curiosity, sometimes tinged with 'concern', about the Indian Navy's (IN) plans, force architecture and future intentions. Presumably this can be ascribed to the fact that, not too long ago, the responsibility for policing the oceans (including our part of the world sweepingly termed 'East of Suez') was neatly parcelled out to a few maritime powers. In this context, let me describe an early lesson that I learnt — in maritime circumspection a quarter of a century ago while on solitary passage from Kochi to Mauritius, in command of a training frigate.

One morning, we sighted an unidentified auxiliary approaching over the horizon in the lonely reaches of the ocean. Partly to provide some training value to the midshipmen, partly to break the monotony but mainly because I felt that an Indian man o' war must do its duty in the Indian Ocean, I sent my crew to 'Action Stations', steered an interception course and challenged the stranger by light and radio. The auxiliary did not care to respond, and as we sighted his ensign at a couple of cables, my overzealous 4.5 inch gun crew got themselves some training value by tracking the passing ship.

By the time I returned home after an enjoyable cruise, our Ministry of External Affairs had received a demarche from the embassy concerned. I received my just deserts in the form of a 'rocket' from Naval HQ for the 'unfriendly and aggressive' conduct of my ship on the high seas!

We have all come a long way since then, and it is a different world altogether today, we meet ships of major navies frequently and exercise with them on a regular basis. Earlier, in the sharply polarized Cold War era, navies collaborated largely within their own political groupings, while those who ploughed a lonely furrow, like the IN, tended to remain insular and even become autarchic. The walls having crumbled, the past decade-and-a-half has seen a dramatic change in international relationships.

The RUSI Defence Systems Journal, Volume 8, No. 2, Autumn 2005.

The globalization of the world economy and the resulting trade and energy interdependence brought home the realization that navies needed to synergize their efforts to ensure tranquility on the high seas for universal benefit. Post 9/11, the common threat of terrorism at sea has brought maritime nations even closer together.

Today's Objectives

Today, India's national objectives lie in ensuring a secure and stable environment, which will insulate us from external intervention and permit social and economic development to proceed unhindered, so that the country can ultimately take its rightful place in the international comity of nations. There is a clear realization that development is just not possible in an environment, which does not guarantee security. There is also clarity that the maritime dimension of the nation's security paradigm has acquired criticality An economically resurgent India has vast and varied maritime interests, which include offshore hydrocarbons, sea-bed resources, sea lines of communication, a vast EEZ, energy lifelines, a substantial merchant marine, ports, harbours and coastal assets.

The Indian Ocean tsunami of 2004 saw 38 ships, 19 aircraft and about 5,500 personnel of the IN being deployed within hours to provide succour, not just to our own countrymen, but also to our neighbours in the Maldives, Sri Lanka and Indonesia in their hour of need. It showed that oceans today are not expanses of water that divide continents, but maritime highways that link nations. This tragedy afforded an opportunity to demonstrate clearly to the world community the ability of navies not just to work alongside each other, but also to catalyze and facilitate cooperation amongst nations. It also taught us in the IN that such contingencies need to be worked into operational plans and catered for while planning force architecture.

As a colony whose markets had to be kept captive for British produce, India completely missed the industrial revolution. At Independence we inherited a motley fleet of two-dozen sloops and minesweepers, which we supplemented later by import – first from the UK and then from the USSR.

It was the urge to attain early self-sufficiency that led to the inception of an ambitious shipbuilding programme, and we commenced construction of Leander class frigates from bought-out designs in Mumbai in the late 1960s. This programme ran into 12 ships but after the first four, which retained the basic design, our architects started improvising. They first 'stretched' the hull to accommodate a Sea King helicopter, and then they 'broadened' the beam to take on two Sea Kings plus a battery of Russian SSMs.

The last ship in this series, the INS *Beas*, commissioned on July 11, 2005 carries a combination of weapons and sensors of Indian, Russian and Israeli origin, seamlessly integrated by our scientists and engineers into a unique and formidable package.

The Long-term Shipbuilding Programme

The recently promulgated Indian Maritime Doctrine visualizes four classical missions for the IN, encompassing military, diplomatic, constabulary and benign roles. The discharge of these missions calls for a range of maritime capabilities, which dictate the force architecture of the IN. Essentially our medium-term plans call for a maritime force built around the core of two operational aircraft carrier groups, which would be capable of exercising sea control in all three dimensions in the distant reaches of the IOR. This force would be supplemented by the requisite submarine and aviation assets, and supported by amphibious, mine counter-measure (MCM) and auxiliary forces. Towards this end, long-term ship and submarine building plans and aviation acquisition programmes have been prepared. From experience, we have derived the lesson that as much as possible of this maritime build-up must be sourced within the country.

Fortunately, over the past 60 years, we have prepared the ground well. India's Defence Research and Development Organisation (DRDO) has a large network of laboratories manned by some of the country's best scientific brains. While a few of their projects have suffered delays, the DRDO, as far as the IN is concerned, has produced 'winners' in the area of sonars (both hull-mounted and towed-array), radars, ESM, ECM, communications, data link and combat management systems; world-class missiles and torpedoes are on the way. India's private and state-owned industries produce ship-building steel, propulsion systems (steam and diesel), gear boxes, gun mounts, torpedo tubes, power generation, air conditioning, hydraulic, pneumatic and a variety of other shipboard systems.

Our force planning process has been interfaced with the country's warship building capability, through the Navy's Controller of Warship Production and his resident Director General of Naval Design. To meet the requirements of the force levels (including two carrier groups) that we hope to field by 2015, there are currently on order 19 ships including destroyers, frigates, ASW corvettes, offshore patrol vessels, LSTs, and fast-attack craft. We already have approvals for another 15-20 ships, but a lot depends on the production rate of our shipyards, and how much private industry can contribute. In extremis, we may have to place a few orders overseas.

India's Carrier Programme

India's carrier building and acquisition programme tends to evoke much interest, and it may be worthwhile offering a few details here. The configuration of the ship was predicated on the fact that a steam catapult (an item built exclusively by the USA) was not going to be available to us, and we would have to use a ski-jump for launching fighters. Having used the ski-jump for many years with the Sea Harrier on the carriers

INS *Vikrant* and *Viraat,* we were comfortable with it, but this choice automatically excluded aircraft like the Hornet, or Rafale and narrowed our options to either the Sukhoi-27 K or MiG-29 K (the suffix 'K' stands for Korabelny or ship borne) – these being the only conventional (non-VSTOL) fighters capable of launch from a ski-jump. For dimensional and weight considerations we have chosen the MiG-29 K.

The former Russian 44,000 ton VTOL carrier, *Admiral Gorshkov* was acquired in 2003 and is currently undergoing modernization and conversion to a new configuration in the northern Russian port of Severdovinsk. This configuration involves installation of a ski-jump forward and a set of arrester wires aft, and is known as STOBAR for 'short take off but arrested landing'.

The keel for a second such ship, the Indigenous Aircraft Carrier (IAC), was laid in Kochi in April 2005, with a completion date of 2012. We did not need to agonize too much (a la CVF), over its configuration because this ship will operate the MiG-29 K too, and therefore have to be STOBAR. At 37,500 tons, the IAC is smaller than the Gorshkov, but as the biggest warship ever built in India, will certainly pose many a challenge to our ship designers and builders. In a parallel technology initiative, our aircraft industry has embarked on a project to design a STOBAR version of the indigenous Light Combat Aircraft, designated LCA (Navy), which we plan to induct on both our carriers by 2015.

With modem sensors, long-range weapons and new ships, submarines and aircraft entering service, we have plans to network our platforms in all three dimensions. We do face technological challenges in several areas, due partly to technology denial regimes imposed on India for many years, but with indigenous effort and foreign collaboration with willing partners we hope to overcome these impediments. The Indo-Russian joint venture, which has produced the supersonic BrahMos SSM is perhaps a good example of things to come. The only question mark over our maritime capability matrix is a submarine building capability, and this is a void, which we hope to address in the very near future.

A Force for Good

India sees her navy as a force for good: assuring peace, tranquility and stability in the IOR and building bridges across the seas, while safeguarding the nation's vital maritime interests. On the Navy also ride many of the country's hopes and aspirations because, of the three armed forces, it is the IN, which has invested most heavily in Indian research and development, technology and industry. We look forward to the future with complete faith and confidence in our indigenous skills and talent.

INDIA'S MARITIME STRATEGY

In October last year, the Indian Navy (IN) promulgated a document entitled, *Freedom of the Seas: India's Maritime Strategy*. This was a classified publication, to which the general public would have had no access. However, since it was desirable that sections of society other than the defence community should be familiar with the elements of the Maritime Strategy, NHQ has decided to issue an unclassified version. So now, both the Maritime Doctrine and the Maritime Strategy will be available to those civil servants, academics and members of the media, who have an interest in such arcane matters.

In the following talk, I have tried my best to ensure that while adhering to the spirit of these documents, I have taken as little as possible from them. Therefore as they say in the titles of many movies: "any similarities that you may note are purely unintended and coincidental."

I would like to start by providing a perspective on the subject of strategy, and why one is required at all. This aspect assumes more importance today, because each Service has promulgated a Doctrine, and we also have a Joint Doctrine. Often the terms "doctrine" and "strategy" get confused, or are even used inter-changeably. So I think it will be useful if I spend some time to provide a distinction between the two.

Distinction between Doctrine & Strategy

According to the dictionary, doctrine is simply "that which is taught", and a NATO definition describes it as "the fundamental principles by which military forces guide their actions in support of objectives." It is also defined as a "framework of principles, practices and procedures, the understanding of which provides a basis for action." It is meant to be authoritative and yet requires judgment in application.

Adapted from the Fifth Annual Major General Samir Sinha Memorial Lecture delivered at the United Services Institution of India, New Delhi, April 30, 2007.

Doctrine is based on the enduring lessons of history and experience, and the repeated success or failure of certain actions over time, tends to elevate them to the level of axioms, which remain relevant to the present and the future. Some of these lessons, deemed through experience as immutable, are placed in an exalted category termed the Principles of War. Of course, doctrine cannot be based exclusively on experience. As Frederick the Great once pointed out, if experience were all that important, he had several mules in his army who had enough experience to be Field Marshals!

On the other hand, a strategy has to be essentially based on a threat; without a coherent threat, whether existing or projected, there may not be a *raison d'etre* for a strategy. As the threat alters its form and shape, the strategy too, must keep evolving and changing in response.

Traditionally, a strategic plan has been associated with the preparation and waging of war. However, since the nature of conflict, the fabric of society, and our geo-political environment has undergone a change post-World War II, the threat has assumed different proportions. Strategy is now, more than merely a military concept since it increasingly requires consideration of non-military matters, with economic, political, sociological and environmental issues driving it into the realm of statecraft.

Strategy can thus be summed up as an overall plan to go from the present situation to some desired goal in the face of a threat, whether in peacetime or a conflict scenario. A strategy will always be set in the context of a given politico-military situation prevailing over a finite time frame and within the ambit of overall national aims.

Doctrine, on the other hand, is a body of thought and a knowledge base, which should underpin the evolution of strategy. Without doctrine, strategists would have to make decisions without points of reference or guidelines.

In this context, the US provides a useful illustration. In pursuit of victory over Nazi Germany, it evolved a strategy for WW II, which required it to strike a Faustian bargain with the Soviet Union and make her an ally. No sooner had the war ended, that the US launched a new strategy of "containment" to prevent a powerful USSR from reshaping the post-war world order according to its own ideology. The strategy to implement containment went through several iterations because the threat from communism kept changing in intensity and geographical focus throughout the Cold War era.

The end of the Cold War brought with it, a complete change in the threat scenario, and the Global War on Terrorism that followed a decade later resulted in a flux in international affairs. All this has caused the US Department of Defence and the US Navy to continuously evolve new strategies and bring out vision documents at the rate of one every four to five years.

Why a Maritime Strategy?

The well-known defence analyst Edward Luttwak asks the question, "What is a Navy in the absence of a strategy? It is, in effect a priesthood." Because, without strategy to guide and inform naval officers, he argues, it is all merely ritual and routine, gold braid and glitter.

In the mid-1980s, I recall reading with great indignation, a statement by the editor of *Jane's Fighting Ships* in one of his Forewords, which said something to this effect: "…the Indian Navy is probably one of the few major navies which first acquires hardware and then thinks about how to use it." In retrospect, I can understand the reason for such a statement, because at that juncture, not only did the Navy lack a doctrine and strategy, but, was truly a "Cinderella" service whose fortunes were hostage to the whims and fancies of the annual budget.

A maritime strategy however, does not concern naval officers alone, and cannot be anything but a sub-set of national strategy. Every nation must have a vision of its place in the world, as well as the role it wishes to play in the international order. Regrettably, in India's case, we have historically suffered from an intellectual vacuum as far as strategic thinking is concerned, and that is why, after 60 years as a sovereign republic we lack a clearly articulated statement of national aims and objectives.

This is a cultural handicap, which has not just deprived us of a healthy tradition of strategic debate and discourse, but also had a deleterious impact on internal security as well as foreign policy issues at the national level. We do not seem to realize that this shortcoming has often been misinterpreted as a sign of weakness and lack of national resolve, and perhaps even acted as a provocation for aggression.

As an aside, I may mention that one of the reasons for this lacuna is India's total obsession with politics, which permits little or no room for other vital issues including national security. In my own experience, the requirements of mere political survival place such heavy demands on a politician's time that even ministers are hard pressed to spare the mental space or attention span for the contemplation of strategic issues.

If I were to state the reasons why the Indian Navy (IN) considered it essential to generate a strategy at this particular juncture, I would sum them up thus:

- Firstly, the sustained induction of hardware from diverse foreign sources that we have seen over the years was never accompanied by any operational expertise or doctrine, because such things are not to be had for money. The hiatus that I just spoke about, has in the past often impacted adversely on doctrine, force planning, equipment acquisition and infrastructure development processes.

- Secondly, while bemoaning the absence of higher strategic direction we have often had to use the Defence Minister's Operational Directive, and even the MoD Annual Report as notional substitutes for a national security document. Thus while the formulation of a maritime strategy was clearly considered overdue by the navy, there was also the possibility that it might provide an incentive for the national security establishment to shake off its inertia, and get to work in this field.

- And finally, India's emergence as a nation of global significance has brought with it the recognition that not just our national security, but also our economic prosperity has deep linkages with the maritime environment.

- At one level, our decision makers are recognizing the key role of the Navy in insulating the nation from external intervention, as well as its vast potential as an instrument of state power. At another level, the realization has also begun to dawn on the intelligentsia that trade and energy, the twin pillars of our economic resurgence, are inextricably linked with maritime power. A clear-cut roadmap is therefore necessary at this juncture to synergise our national maritime endeavours.

India's Geo-strategic Environment

The Indian peninsula juts out over 1,000 miles into the Indian Ocean, and her geo-physical configuration makes her as dependant on the seas as any island nation. This predicates the profound influence that this ocean, the only one named after a country, will have on India's security environment. K.M. Panikkar summed it up neatly, half a century ago, in these words: "While to other countries the Indian Ocean is only one of the important oceanic areas, to India it is a vital sea. Her lifelines are concentrated in that area; her freedom is dependant on the freedom of the sea-lanes."

The Indian Ocean Region or IOR, at whose focal point India is located, has some unique general features, of which we must take note before examining specific issues.

- Amongst 56 nations of the IOR, some of the fastest growing economies of the world co-exist with some of the poorest. Many of the countries are afflicted with serious problems of backwardness, fundamentalism and insurgency. Most of them are under military dictatorship or authoritarian rule.

- This region is the largest repository of the world's hydrocarbon resources, and apart from producing the most rubber, tin, tea and jute, is well endowed with strategic materials like uranium, tungsten, cobalt, gold and diamonds.

- The region is home to one-third of the world's population which is regularly struck by 70 per cent of the planet's natural disasters.

- Most of the post-Cold War conflicts have taken place in this region, and today, the global epicenter of terrorism as well as nuclear and missile proliferation exists right next door to us.

- Areas of the IOR, like the Horn of Africa and the Malacca Straits are rife with incidents of piracy, gunrunning, drug-trafficking and hijacking.

Territorial and maritime boundary disputes, runaway population growth, and the migrant labour economy of South Asia are some of the other generic factors which need to be noted, as containing the seeds of future conflict. As far as the regional and other players are concerned, we need to spend a few minutes to make a brief assessment of their impact, current and future, on the region.

First, a look at the sole superpower USA, which has to be counted as a regional player by virtue of her large and ubiquitous maritime presence in the IOR.

- It is clear that for the foreseeable future, the US will continue to remain deeply engaged in the IOR and the wider Asia-Pacific region. The two abiding US interests in this region are: safeguarding the hydrocarbon resources of the Middle East and Central Asia, and the containment of China to protect the autonomy of Taiwan. The Asia-Pacific geographic area has been divided along the Indo-Pakistan border, between the Hawaii-based Pacific Command and the Florida-based Central Command.

- Currently, America's resources and attention are intensely focused on the ongoing operations in Iraq, and the requirement to keep the terrorist hubs in Pakistan and Afghanistan under check. Monitoring, and finding ways to circumscribe the nuclear ambitions of North Korea and Iran, are the other two issues that have critical significance for the USA today.

- It is now becoming obvious that while she can try and set an agenda to suit her interests, the USA cannot by herself, implement such an agenda worldwide. Signs of "imperial overstretch" are surfacing, as are low key feelers seeking help and support, especially at sea. Under these circumstances, a helping hand from a respected regional power like India would bring welcome relief. The stage for this has been set by the passage of the Henry Hyde Act and rapidly warming relations between the armed forces.

Next, let us look at China, which though not on the littoral, looms menacingly over the IOR as a rapidly emerging entity with her sights set firmly on super-power status. In the context under discussion, there are just four major points to be noted.

- Firstly, China's nearest competitor in both the military as well as economic spheres is India. Since both are Asian powers, it is a historical inevitability that they will have to compete and even clash for the same strategic space.

- Secondly, with Sino-Indian bilateral trade having crossed the US$ 20 billion mark, China is well on the way to becoming our largest trading partner. This is a welcome development, but which also contains a contradiction. We must not allow it to lull us into a sense of complacency, because the Chinese have not forgotten our territorial disputes. Just a fortnight before the visit of President Hu Jintao in November 2006, in a most undiplomatic gesture, the Chinese Ambassador in New Delhi reiterated an emphatic public claim to Arunachal Pradesh. It is significant that China has settled boundary disputes with 14 out of her 15 neighbours; the only exception being India.

- Thirdly, the "string of pearls" strategy is another source of concern, due to the clear connotation of military encirclement that it conveys to India. In this context, Gwadar, situated at the mouth of the Persian Gulf is probably the first in a chain of ports that China is helping our neighbours to develop, and which could provide future facilities to the PLA Navy ships and nuclear submarines. The other ports in this chain are likely to be: Hambantota in Sri Lanka, Chittagong in Bangladesh and Sittwe in Myanmar.

- And finally to all those who get dreamy-eyed about the future of Sino-Indian relations, I would put just one question. Where in the annals of international relations can one find a precedent for one nation handing over to another, not just the designs and expertise, but also actual hardware relating to nuclear weapons and a family of ballistic missiles? Even the British were denied atomic secrets by their Anglo-Saxon cousins, the Americans, using the post-War McMahon Act.

- By arming Pakistan with conventional and nuclear weaponry, China has, by proxy, forced India to divert scarce resources and thus tried to checkmate her as a military and economic rival.

Coming to our immediate neighbourhood, India's attitude of detachment with regard to most geo-political developments is often worrisome. Unless we are involved, we will have no leverage, and unless we have some leverage, we are powerless to influence the course of events vital to our national security. In this context, two examples are illuminating.

- India's national security interests have suffered the most from the sinister nexus between China, Pakistan, and North Korea, to accomplish nuclear and missile proliferation, much of which has taken place by the sea route. The Proliferation Security Initiative (PSI) was meant for the specific purpose of interdicting transportation of WMDs by ship, but we have yet to make up our mind about joining it.

- Nepal, Sri Lanka and Bangladesh in our close volatile neighbourhood, are countries where we have chosen to remain hands-off, but where things could blow up in our face overnight, and catch us unprepared, because we have no "trip-wires" in place.

Next, a brief look at Pakistan. While Pakistan will remain a factor in our security calculus for the foreseeable future, we need to be careful that this troublesome neighbour does not dominate our radar. It is certainly in our interest that she should remain a stable and integral nation, and outgrow the sense of insecurity which has haunted her since independence. Our national security policy should convey a couple of clear messages to the Pakistani leadership, such as it is.

- Firstly that India has vast resources of strength, and has demonstrated the resilience to withstand whatever Pakistan could throw at us. We will therefore never be cowed down either by force, or by terrorism.

- Notwithstanding threats of a "thousand cuts", India will continue her march on the road to economic, industrial and scientific progress with resolve. Engaging in an arms race with India could break the Pakistani economy's back.

- And finally, modern day governments need to focus energies on providing a better quality of life to their citizens through social change and economic development, rather than by breeding fundamentalism in madrassas.

Finally, we need to bear in mind that the drug traffic emerging from the Golden Triangle and the Golden Crescent on either side of India provides funding for international terrorism. Organizations like the al Qaeda and the Jemmah Islamiah find both recruits and financial sustenance here, and use the sea routes for their nefarious activities. The LTTE not only has a "Sea Tiger" wing, but also runs a clandestine merchant fleet, which provides efficient logistic support for its insurgency. In addition, the Horn of Africa, Bay of Bengal and the Malacca Straits are witness to frequent incidents of lawlessness including piracy, hijacking and human trafficking.

In the midst of such a scenario, the smaller island nations of the IOR are beset by a feeling of insecurity and seek reassurance from neighbouring maritime powers that their sovereignty will remain protected, and that they will receive succour in times of need.

The last word in any discussion on geo-politics must go to Lord Palmerston who so rightly reminded us that in international relations, core national interests always take precedence over sentiments like friendship or animosity. As a corollary, it must always remain etched in our minds that should a clash of interests arise between India and any other power, regional or extra-regional; the use of coercive power and even conflict remains a distinct possibility.

Security of Energy and Trade

India's own dependence on the seas, and her geographic location astride major shipping routes of the world place a dual responsibility on her. Not only does India have to safeguard the maritime interests vital to her own security and economic well-being, but, she must as an obligation to the larger world community, ensure the free flow of vital hydrocarbons and commerce through the IOR sea lanes.

India, with a merchant fleet of 760 ships totaling 8.6 million tons GRT, ranks 15th amongst seafaring nations. This fleet, operating out of 12 major and 184 minor Indian ports can carry a little less than a sixth of our seaborne trade, and has much scope for expansion. As far as our foreign trade is concerned, I reiterate three oft quoted facts to emphasize the role of the sea:

- Of our foreign trade, merchant ships carry more than 75 per cent by value, and over 97 per cent by volume.

- Our exports were US $ 100 billion in 2006, and are slated to cross US $ 200 billion in the next five years.

- India's share of total world trade has been hovering around just one percent, and the government is aiming to double it by 2009.

Currently at a very energy-intensive state of its development, India is predicted to become the world's largest importer of hydrocarbons by 2050. A new development in this context is our acquisition of oil and gas fields across the globe by ONGC Videsh. While investments worth billions of dollars have been made in these overseas assets extending from Sakhalin across Africa to Brazil, little thought has been given to their protection, which will certainly have maritime security connotations.

Above and beyond whatever our offshore oilfields are currently producing, the seas hold out tremendous promise in terms of oil, gas, and mineral seabed wealth, as well as organic resources. We need to therefore not only safeguard our vast exclusive economic zone (EEZ), but also mobilize the technologies required to exploit these resources.

Annually, over 100,000 merchantmen including bulk carriers, container ships, car ferries, oil tankers, and liquefied gas carriers transit the waters of the Indian Ocean, carrying cargo worth about a trillion dollars. Both East bound and West bound shipping has to pass through a number of choke points where it is vulnerable to interdiction or interference by state and non-state entities. Any disruption in the supply of energy, or commodities, would send prices skyrocketing, and destabilize industries as well as economies worldwide.

It is against this backdrop that India as a major maritime power in the IOR has to shoulder responsibilities. Let us then look at how our maritime strategy envisages the employment of national assets in peace and in war.

Strategy in Peacetime

Let me quote a passage from the first version of the US Maritime Strategy promulgated by the Chief of Naval Operations Admiral James Watkins in 1986. It says: "Sea power is said to be relevant across the spectrum of conflict, from routine operations in peacetime to the provision of the most survivable component of a nation's forces for deterring strategic nuclear war. The maritime strategy provides a framework for considering all uses of maritime power. Amongst the greatest services we can provide the nation is to operate in peacetime and in crises in a way that will deter war."

Our Maritime Doctrine has clearly outlined the four basic missions assigned to the Indian Navy, which span the full spectrum of conflict, and these are: Military, Diplomatic, Constabulary, and Benign roles. Of these, the diplomatic, constabulary, and benign roles are undertaken in peacetime, and shall receive priority in discussion. However, from the wartime military role, I will de-link deterrence, which happens to be a peacetime function, and discuss it first of all.

Strategic Deterrence

Having crossed the nuclear Rubicon in 1998, we are committed to maintaining a minimum credible deterrent under a self-imposed condition of No First Use (NFU). Having also gone public with a nuclear doctrine, there is no room for second thoughts now, because our adversaries have designed their strategic forces, and crafted their nuclear posture based on our declarations.

India's Nuclear Doctrine clearly envisages, and is based on a deterrent in the form of a "triad" with land-based, aircraft-borne, and submarine launched weapons. Of this triad, we only have the first two in our inventory at present. Nuclear weapons are not meant for war fighting, and achieve deterrence by convincing the enemy of the futility of a nuclear first strike, because the response would be so devastating that it would render the strike meaningless.

Two specific attributes are necessary to convince the enemy of the credibility of your deterrent; it should have massive destructive power, and it must be substantially survivable in the face of a sneak first strike. The only platform, which can claim to be virtually invulnerable to attack and ready for instant response is the nuclear propelled submarine armed with strategic weapons. We do have ship-launched ballistic missiles in our arsenal, but our nuclear deterrent would attain true credibility only when its underwater leg becomes operational.

However, operating a submarine-based deterrent is a complex undertaking, and requires not only fail-safe communications, but also a full-fledged command and control system, backed up by extensive intelligence, planning, training and maintenance infrastructure. We did operate a Charlie I class nuclear submarine on a lease from the USSR for three years, and that provided useful exposure to navy personnel and perhaps scientists.

But INS *Chakra* as she was called, carried no nuclear weapons, and we are therefore, still to learn the complexities of operating a seaborne deterrent. If we are looking forward to deploying such a weapon system in the next few years, perhaps now is the time to start putting the components in place.

The unique doctrine of NFU, does raise the threshold of a nuclear conflict, but requires us to maintain adequate conventional forces in addition to our strategic assets. And that brings us to the issue of conventional deterrence.

Conventional Deterrence

As I just said, nuclear weapons are not meant for the purpose of fighting wars, and every soldier, sailor and airman would do well to remember that these are actually political tools to be used only for sending a message of deterrence, compellance, or coercion to an adversary. But such a situation should arise only when persuasion or dissuasion by all other means has been tried without success and herein lies the need for conventional deterrence.

The main objective of a peacetime strategy is to deter conflict, and ensure peace and stability in our areas of interest. This is best done by maintaining a preponderance in maritime capability; the idea being to never leave friends as well as adversaries or potential adversaries in doubt about India's superiority at sea.

In this context it is important to convey a clear and unambiguous message to all concerned, so that there is no room for misunderstandings. A message to convey reassurance or deterrence can have many nuances, and I shall just mention the three main elements:

- First of all, comes what we now call maritime domain awareness or MDA. It is encompasses the ability to keep our oceanic areas under sustained surveillance so that we can compile a comprehensive picture of the peacetime merchant traffic as well as deployment and operating patterns of naval forces. The availability of such a picture not only reduces the "fog of war" but also gives us an early warning of any deviations from the normal state. MDA requires many input sources; from satellite surveillance, aerial reconnaissance, and scouting by warships, to D/F stations and merchant ship tracking systems. Making this information available in real time to all our widely dispersed platforms at sea will call for networking of a sophisticated nature.

- The second element involves our ability to ensure presence, or physically position units in an area of interest, no matter how distant, and sustain them for as long as necessary. Our vision encompasses an arc extending from the Persian Gulf to the Malacca Straits as India's legitimate area of interest. The presence of our maritime units not only enhances our familiarity with those waters, and boosts intelligence gathering, but also demonstrates our commitment to friends in that area, and willingness to protect our interests.

- The third and most important element of this message that would reinforce deterrence is one of combat efficiency and readiness. While we may consider these as inherent attributes of a professional maritime force, overt demonstrations through overseas deployments, joint exercises, training exchanges and even humanitarian relief operations are keenly observed and noted by friends and rivals alike.

Diplomatic Role

"Gunboat diplomacy", used to be considered as one of the less pleasant coercive tactic used by colonial powers in the heyday of imperialism. However, maritime diplomacy obviously has no such connotation, because navies are now being increasingly used to build bridges, to foster mutual trust and confidence, to create partnerships through inter-operability and to render assistance if required.

For our maritime strategy, this role is has two-fold importance. Firstly, the Navy can discharge its peacetime functions in a far more efficient manner, if we have an atmosphere of cooperation with our neighbours and we have friendly ports and airfields available to our units. Secondly, in times of crisis or war too, operations proceed much more smoothly if the environment has already been shaped, and a certain level of comfort established. This aspect emerged clearly during both the Tsunami relief and the Lebanon refugee evacuation operations.

However, this role would acquire true significance only if it forms an integral part of the nation's overall diplomatic initiatives, and the political establishment as well as the MEA learn how to use the Navy as an instrument of state policy. On its part, the IN has accorded the highest priority to what it calls "International Maritime Cooperation" and has created structures, schemes and financial heads to pursue it vigorously. Friendly IOR neighbours, on their part have offered us refueling and re-supply facilities in a number of ports.

Constabulary Role

The constabulary role in our context, must be seen in two dimensions; ensuring good order at sea and managing low intensity maritime operations (LIMO).

A combination of factors and an unfortunate convergence of interests, make the high seas a fertile ground for criminal organizations and terrorist activities. So when we speak of threats to "good order" at sea, it encompasses the full gamut of lawlessness: from piracy, hijacking and trafficking in arms, drugs and humans to smuggling of WMDs. In our own context, the effective sealing of land routes from Pakistan has forced clandestine traffic into the sea, and opened a new window of vulnerability all along our western seaboard.

Therefore, while good order at sea is certainly an internal security imperative for India, it is also an important bilateral and international maritime obligation.

LIMO involves countering non-state entities using the sea for unlawful purposes or indulging in violent activities against states. In this context, we have a tangible threat from our west where, terrorism breeds unfettered with state support, and are likely to spill over into the sea. To the south, the Sea Tiger wing of the LTTE, which has attained proficiency in maritime operations, operates at our doorstep.

The Service has to tread with a degree of caution in the constabulary role because it is very easy to become excessively involved in low end or "brown water" operations, to the detriment of the Navy's prime tasks, which lie in the "blue waters" or the high seas. With the creation of the Indian Coast Guard in 1978, most law enforcement aspects of the constabulary role within the Maritime Zones of India have been assigned to them. The Navy needs to stand behind its sister maritime Service and render support and assistance when required.

Benign Role

The Navy is the repository of certain capabilities and specialist knowledge which are instrumental in the discharge of its benign role. This role involves tasks such as humanitarian aid, disaster relief, search and rescue, diving assistance, salvage and hydrographic surveys and is essentially defined by the complete absence of force or violence in its execution.

The tsunami of December 2004 provided an example of how the basic attributes of maritime power enable it to react at short notice, and respond to emergent situations. The disaster struck on a Sunday morning, and as our ships were sailing for our own east coast and the Andaman Islands, we received requests for help from Sri Lanka and Maldives. We mobilized more ships, helicopters, medicines and stores, and by the same evening they were on their way to Male and Galle. Government approval came later by phone, but we knew that if there had been a problem, our ships would just anchor 12½ miles offshore and await instructions.

Similarly in June 2006, our ships were returning from the Mediterranean when the Lebanon refugee crisis arose. We ordered them to anchor in the Suez Canal while the MEA pondered over the issues involved. As soon as the Government decision was received, they turned around and sailed into Beirut to commence the evacuation operations within hours.

These two operations have had a significant impact, and served to enhance India's image in the international community. It is to be hoped that the establishment has drawn the right conclusions about the employment of the navy as an instrument of diplomacy.

Strategy in War

I have dwelt at some length on the Navy's peacetime strategy because peace fortunately prevails about 90 per cent of the time. But we have to remember that the prevalence of peace is an indicator that deterrence is working. Should deterrence fail, war will surely follow, and war is what navies train and prepare for.

An essential element of this preparation for war is the evolution of a new maritime strategy. Apart from the other imperatives that we have discussed earlier, this evolutionary process has been accelerated by economic, geo-political and technological developments that have come about in the recent past.

Before embarking on a discussion of the strategy, I would like to make two important points, which may call for a paradigm shift.

- Firstly; under the influence of Mahanian ideas, most navies including our own, imagined that their *raison d'etre* was only to engage the enemy in a big battle at sea, and plans were shaped accordingly. However, the lessons that emerged from exercise after exercise clearly conveyed that navies cannot achieve a great deal, conducting maritime operations in isolation. Unless our actions at sea had a linkage, no matter how indirect, with events on land, the navy's potential would be wasted. There is now acknowledgement that wars are won only on land and that the Navy must ensure that its planning process as well as operations, are synchronous with those of the army, so that we obtain the maximum synergy.

- Secondly; there is a section of opinion, especially in the army and air force, which firmly believes that all future wars in our context, should be "short and sharp". Perhaps it is a Hobson's choice for these Services because the intensity of FOL and ammunition consumption as well as attrition can be limiting factors for them. As far as the Navy is concerned, the longer a conflict lasts, the greater the pressure that it can bring to bear on the enemy. As the Vietnam, Iran-Iraq, Kosovo, and current Iraq wars have shown, short conflicts are not an inevitability and we should retain the option to prolong a conflict if it suits our national interests.

Maritime forces can be deployed in two ways to influence the outcome of war on land. They can be used to interdict the enemy's foreign trade lifeline in an attempt to starve his industry, economy and people, and bring his military machine to a halt. The impact of this "commodity denial" or "indirect" regime requires a finite time to be felt by a nation. Factors like the enemy's dependence on imports, his buffer stocks and ability to re-stock via land routes will decide the effectiveness of these indirect operations, and that is why a superior navy would like to prolong a war.

In the other, "direct" mode of creating an impact on the land battle, the enemy's homeland is targeted by naval platforms delivering weapons from the sea, undertaking amphibious operations or inserting Special Forces. With the demise of the Soviet Union, the open-ocean warfare challenge disappeared, and the USN-Marine Corps combine shifted their focus to crisis-response and interventions in the third world. Herein lies the origin of concepts like "littoral warfare" and "naval expeditionary forces".

Adapting these concepts to our environment, the maritime strategy must encompass the resolute and judicious deployment of our maritime forces in both direct and indirect operations. This will ensure that the impact of sea power is felt on the land battle, both in the short term and long term time frames.

Neither littoral warfare nor expeditionary warfare are new functions, but essentially involve a geographic relocation of the theatre from mid-ocean to a zone extending about 50-100 miles inland and seaward from the enemy coast of interest. All the other traditional forms of naval warfare, like amphibious anti-submarine, anti-aircraft, electronic and mine warfare would retain their importance. However, there are some concepts and factors, mostly technology based, that we would need to incorporate into our new strategy.

- The littoral of an adversary is an inherently dangerous area for maritime operations due to threats from submarines, strike aircraft and mines, etc. Therefore it would be essential to impose a sequencing of operations so that the battle space is adequately sanitized and favourable conditions created prior to launching any operations.

- In such sequencing or phasing, it would be imperative to first establish information dominance in order to disrupt the enemy's command and control systems and deny him information about our intentions. Thereafter, sea control, a favourable air situation, or mine counter-measures as appropriate, could be pursued before the actual operation is launched.

- Although a new buzz word, all that information dominance, means is attaining superiority in the electromagnetic as well as information warfare domains for one's own forces while destroying, degrading and even deceiving the enemy's intelligence and surveillance assets. We should have no doubts that this would be a decisive factor in any future conflict.

- Today, our fleets possess tremendous striking power in terms of number of SSMs, ASMs and SAMs that can be launched from our ships, submarines, aircraft and helicopters. However, this punch would be wasted in a conflict unless we can bring the enemy to battle. Our forces would therefore have to aggressively seek out enemy units and bring them to action so that we can inflict adequate attrition prior to attacking his homeland.

- In order to obtain the maximum synergy and advantage from our superior numbers as well as capabilities, it is necessary that we fight what the Soviets used to call a "combined arms battle" at sea. By ensuring reliable and secure communications, between warships, aircraft and submarines, it should be possible to concentrate their firepower in a geographical location and inflict heavy attrition on the enemy. Shore based IAF strike aircraft would make an important contribution here. With network centric operations on the horizon, implementation of this concept should not pose a problem.

- Combined arms operations fit neatly into another concept termed: "maritime manoeuver from the sea". Given their inherent mobility and the access provided by the sea, maritime forces can exploit the principles of Surprise, Concentration, and Flexibility to deal the enemy a sudden blow which will unbalance him and shatter his morale and cohesion. Essential ingredients for such an operation include naval aviation, land attack missiles, amphibious shipping and Special Forces. These are all available to the IN, and manoeuver warfare should be an important part of our strategy.

- In the final phase, our strategy should envisage the linking up of the three Services in a joint operation, no matter how widely dispersed these forces, or brief this phase may be.

Ideally speaking, maritime force structures should evolve from an approved strategy. But having made a late start in this domain, we will have to make some compromises till the cyclic process in which strategy leads to capability requirements, which in turn influence the force planning process, stabilizes.

Nevertheless, the IN has not done too badly; having generated a Doctrine, a Maritime Capabilities Perspective Plan, and a Strategy within a span of two years. The thought process and discussions that went into the evolution of these documents has generated a need-based, budget-linked force structure for the next 15 years, which has been accepted in principle by the MoD.

Epilogue

It is not entirely happenstance that the Navy as it evolves, will meet most of the demands of India's Maritime Strategy over the next decade and a half. It did not happen overnight, and a great deal of credit for this should rightly go to our farsighted predecessors who laid sound foundations and put the Service on the right track

I must make one mention of one last set of issues, before closing. Just as Strategy forms the basis of operational plans, it must itself be supported by a philosophic underpinning, which will help the Navy retain a clear vision of the future, and steer a steady course. This underpinning is provided by a set of five factors which I would commend to the Navy for close attention.

- **Indigenization:** India today has the dubious distinction of being the largest arms importer in the world, having signed deals worth US$ 11.7 billion over the past two years. Experience has shown us that every time we induct a system of foreign origin, we are entering a dangerous cycle of spiraling costs, uncertainty, and dependence on an unreliable supply source. Self-reliance should remain a key result area and, for all their clumsy ways, we should continue our symbiotic relationship with the DRDO. Firmly rejecting "screwdriver technology" we should insist on the DRDO entering collaborative development, and co-production arrangements wherever we are offered transfer of technology.

- **Shipbuilding Industry:** Our current status as a maritime power is due in substantial measure to the 40 years of warship building endeavours of our shipyards. The shipbuilding industry is a strategic asset, which must be carefully nurtured and guided by the Navy. Apart from undertaking urgent modernization, the shipyards must be encouraged to seek partnerships with the private sector and technical collaborations abroad.

- **Foreign Cooperation:** The Navy's most important contribution to the nation during peacetime is going to be as an instrument of diplomacy, providing support for political objectives and foreign policy initiatives. In coordination with Ministry of External Affairs a sharp focus will have to be retained on coordinating assistance to our maritime neighbours in the Indian Ocean littoral in areas of training, hardware and expertise.

- **Networked Operations:** Our maritime forces currently encompass weapons, sensors and platforms of formidable range and capability. With the induction of the aircraft carrier *Vikramaditya*, systems like the Brahmos missile and new classes of submarines, our capabilities at sea will be further enhanced. In order to exploit their full potential, we will need to have a sophisticated network covering the entire IOR. With a dedicated maritime communication satellite and the help of our IT industry we should aim to have a world class network in place by the middle of the next decade.

- **Transformation:** Change of any kind does not come easily to us, because we dislike the associated turbulence, and dread the thought of failure. But the choices are stark; we either look ahead and bring about an orderly sequence of change through "transformation", or get overtaken by events and react to them *post facto*. Transformation is the engine, which will help the Service absorb new technologies, move towards networked operations, make organizational improvements, embrace joint philosophies, and incorporate other ideas to improve combat efficiency.

MARITIME TECHNOLOGY AND SHIPBUILDING

INDIA'S FIRST INDIGENOUS AIRCRAFT CARRIER
A Brave Venture

The "steel cutting" ceremony which the Minister of Shipping and Transport, T.R. Baalu, has so kindly consented to grace, marks a very special moment of pride, joy and emotion for all personnel of the Indian Navy, serving as well as retired. I have no doubt that this day will be seen as a significant milestone in our maritime history by generations of Indians which follow. I, therefore, feel specially privileged to be present here in Cochin Shipyard to witness the historic inauguration of work on our first Indigenous Aircraft Carrier, by a very distinguished personality.

By a happy coincidence, today's ceremony will be the second major maritime event to take place in Kerala this month. The first was the commissioning of INS *Zamorin* at Ezhimala by the Chief Minister on April 6. The State of Kerala, with its rich maritime tradition has always rendered staunch support to the Indian Navy, and has thus exerted a vital and positive influence on the growth of our Service. Each one of us has spent our formative years in Kochi, and we have a strong emotional bond with this hospitable city.

Speaking of Kerala's maritime history, one must never forget that a defining moment in India's history occurred in this very state. Five hundred years ago, on the May 20, 1498 the Portuguese adventurer Vasco da Gama landed in the port of Calicut, and sought trading rights from the Zamorin. This marked the beginning of European influx into the land of our forefathers, because the Portuguese were followed by the Dutch, the French and the British who came looking for spices and gold, but stayed on to exploit India.

India's enslavement by foreign powers until our independence in 1947 was a direct result of a lack of appreciation of the importance of sea power, and consequent neglect of the seas. Except for a few brave patriots like Kunjali Marakkar and the Angres, there was little resistance to the exploitation of our seas by outside powers. The fact that final domination of India by alien powers resulted not by overland invasion but by an onslaught

Adapted from a speech at the Cochin Shipyard, on the occasion of the steel-cutting ceremony for India's First Indigenous Aircraft Carrier, Kochi, April 11, 2005.

across the seas is a fact that is now indelibly imprinted in the mind of every Indian.

With this realization came a resolve to re-vitalise India's glorious maritime tradition of shipbuilding. If the Wadias of Bombay could build sound sailing ships for the Royal Navy in the 18th and 19th centuries, why could their successors not uphold this tradition for the Indian Navy in the 20th century? We made a brave and pioneering start in the late 1960s in Mazagon Dock with the *Leander* class frigates, whose design was based on Royal Navy ships of the same class.

The Indian Navy thereafter graduated, first to modifying existing designs, as with *Godavari* class frigates, and later to entirely new designs such as the graceful and potent *Delhi* class of destroyers. Today, we have four or five different classes of warships at various stages of construction in our shipyards.

From destroyer to aircraft carrier is however a major step. The design and construction of such a ship will mark the "coming of age" of our indigenous warship-building capabilities, as this will be the largest as well as the most complex warship to be ever built in India.

The present century is going to be an oceanic one and India has a significant stake in the seas that surround us on three sides. We are already heavily dependent on the seas for food, energy and minerals. 68 per cent of India's oil production comes from offshore resources and 70 per cent of our crude is imported – all of it over the sea. The new 'Silk Route' originates from the Persian Gulf and passes through the Indian Ocean, around the Indian Peninsula through the Malacca Strait to the Pacific Rim countries. Over 60,000 ships are known to transit through the Indian Ocean every year, transporting oil, consumer goods, food and electronic goods worth an estimated US$ 1,800 billion.

Consequently, the oceanic area of direct interest to us extends from the Persian Gulf, down to the east coast of Africa, and across to the Malacca Strait. A blue-water navy, capable of protecting our far-flung maritime interests in this vast ocean region is therefore absolutely essential for a nation of India's stature.

The indigenous aircraft carrier construction programme is a critical component of this blue-water force, to provide sea control and power projection capabilities to the Indian Navy. As we cut steel for this ship, we join an elite club of just six other nations, which are able to conceptualize and then realize an endeavour of such dimensions and complexity. The aircraft carrier is expected to enter Service in 2012 and will have a life span of 50 years. Its capabilities will span all three dimensions and its potent embarked airpower will tilt the balance at sea.

To the management and workers of Cochin Shipyard, I would like to say that by entrusting the construction of the prestigious indigenous aircraft carrier to this shipyard, the Indian Navy and indeed the country has reposed great faith and trust in your capabilities and dedications. Knowing the fine reputation, good work, ethos and excellent past record of this shipyard, I am confident that you will turn out a high quality ship on time for the Navy in the best traditions of CSL.

Network Centric Operations
The Need for Shared Awareness

Network Centric Operations, or NCO, relies on computer processing power and networked communications technology to provide a shared awareness of the battle space. This shared awareness increases synergy for command and control, resulting in superior decision-making, and the ability to coordinate complex military operations over long distances, with the ultimate aim of obtaining an overwhelming military advantage. In fact, NCO is a key component of the 'Transformation' that is taking place in advanced militaries around the world.

Transformation is happening in almost all fields of human endeavour. Defence transformation involves large-scale, discontinuous, and possible disruptive changes in military weapons, organisation and concept of operations that are prompted by new and significant changes in technology or the emergence of new and different international security challenges. Improvements in computer technology, with their resultant impact on computing power, communications and the lethality of weapons are one of the main drivers of transformation. It would interest some of you to know that the Navy has set up a separate directorate to look at the issue of transformation and what we need to do about it.

What has been discussed here is the management of information – seamlessly, in real-time and unencumbered by unnecessary detail. The aim, of course being, to eliminate or reduce the "fog of war" and achieve not just information superiority, but more importantly, decision superiority.

The relationship between information and combat is well known. However, the challenge has always been to see how it can be maximised. All military operations are conducted

Adapted from the Valedictory Address delivered at a seminar on NCO jointly organised by the Centre for Land Warfare Studies, the National Maritime Foundation and the Centre for Airpower Studies, New Delhi, December 21, 2005.

in three domains. Two of these - the physical domain and the domain of the mind are well known and understood. The third domain is that of information. It is this domain, which is now being exploited to increase combat power in a broad range of operations. This is extremely important as future conflicts are expected to be short, fast paced and intense. Hence the availability of timely and continuous information, which results in speedy and correct decisions, will magnify the effectiveness and efficiency of a military force several times over.

Let me talk about my personal experience in dealing with this very important issue and what, in my viewpoint, needs to be done to progress it in the Indian Armed Forces.

The first issue is awareness about NCO itself. While most people know what NCO is in general terms, there is insufficient awareness about its nuts and bolts. In the absence of this knowledge, decision-making becomes extremely difficult and convoluted. There is no harm in admitting to this fact as most militaries, perhaps with the exception of the United States, are in the same boat. Therefore, the first and most important task is to spread awareness about NCO. At the same time, we are fortunate to have a core of officers who have a deep understanding of this subject and we need to give them sufficient freedom and initiative to get on with the task at hand.

The second issue is related to the first and deals with methodology of introducing this concept in the Indian Armed Forces. In my opinion, there is no need to reinvent the wheel because of the speed at which computer-related technologies are advancing the world-over. There are sources - both within the country and abroad - from where the basic concept and architecture of NCO can be taken. Of course, there are some sensitive areas where security considerations will dictate an in-house approach. But in general, particularly as far as software solutions are concerned, commercially available solutions and out-sourcing is the recommended way to go, in order to leapfrog our way ahead.

The third issue relates to the need to look at NCO in a holistic manner. Network-centric operations will impact at all levels of war - strategic, operational art and tactical. As I mentioned earlier, NCO will necessitate changes in not only our equipment, but also in the way we train our personnel, the way we intend to fight our future battles and indeed in the way we are organised. These all-encompassing changes need to be recognised up-front and the big picture seen *in toto*, if we are not to replicate the fable of the "blind men and the elephant".

Fourthly, NCO involves, not just the introduction of new concepts and software and hardware, but also a new way of thinking. This is because the democracy of awareness that a networked environment will provide, will also allow lower formation commanders to take informed decisions. It will also allow superior commanders to intercede at levels that were not possible earlier. This will call for greater flexibility of thought and operations, as the traditional rigid, vertical hierarchical decision-making process will give way to a more multidirectional, dynamic and inclusive process. While this will deliver a range of benefits, it needs to be noted that collaborative and devolved decision--making via networks may cause information overload, command gridlocks and even a degree of chaos in operations. Networks are complex systems that, unlike hierarchies, thrive on connectivity, flattened command structures and 'peer-to-peer' nodes of communications. In a long chain of command dependency, small failures can echo across a network. As network complexity increases, solutions to problems in one node are likely to require parallel adjustments to behaviour in other nodes.

The fifth and final issue relates to jointness. NCO will only truly be effective if it results in a tri-Service synergy. This is well recognised but what needs to be stressed is that since each Service fights in a different environment, the software and hardware requirements will be unique. Hence each Service needs to proceed along an independent path while simultaneously ensuring that their systems and software are capable of talking to each other through a set of common protocols. This will be the responsibility of HQ IDS and they are already engaged in this direction. A joint C4I2 Doctrine has already been approved and it to be promulgated shortly.

Individual Services tend to design their future operations around current (or slightly modified) forces, with an overlay of network improvements, based on today's knowledge. It needs to be remembered that future networks may create an operational environment that cannot be anticipated or predetermined. The professional military inclination is to seek to control change, since control means order and predictability. It is a military characteristic to try to plan, to the last detail, the evolution of our networked force. Yet, it is almost certain that the consequences of networking the force will contain unpredictable features. In the implementation of the road map we should seek general direction, but we must possess the imagination to exploit unforeseen opportunities as they arise.

By way of conclusion let me say that all new technologies represent a challenge to an existing social order and imply gain for some constituencies and loss for others. There is also an intra-organisational competition for resources and status. Since the results of networked technologies are likely to have their greatest impact on the sociology of military organisations, the greatest challenges that the Armed Forces will face are likely to be cultural, in the form of introducing changes in thinking and behaviour. Hence, existing attitudes and beliefs about how warfare is conducted today may well be the biggest impediment in achieving network-centric capabilities. In order to overcome this, we must remember that increase in information sharing has the potential to create dramatic improvements in both single Service and joint war fighting capabilities, to our collective benefit. Hence, we need to grasp this opportunity jointly, and there is no time to lose.

Transformational Technologies for Navy of The Future

Singapore is a unique city-state of our times, which stands as a beacon for democracy, order and stability in this region. I am pleased to be here as a guest of the Republic of Singapore Navy, whose Chief, Admiral Ronnie Tay is present here. Our two services share a close relationship of long standing, and over the years, we have developed mutual confidence and a great respect for each other's professional abilities and attainments.

Navies have a deserved reputation for being extremely tradition-bound and conservative in their outlook. Historically, the "silent Service" has kept aloof from radical changes of all kinds; one exception being the Russian cruiser *Aurora*, which fired the opening shot of the October Revolution, in St. Petersburg. Navies treated with extreme caution even, technological change. The steam engine, the torpedo, the airplane and many other innovations had to fight entrenched suspicion and scepticism to find their place in navies. Such conservatism led a senior Royal Navy officer at the turn of the century to condemn the submarine in these words: "...underhand and damned un-English. Treat all submarine crews as pirates and hang them!"

However, it must be admitted that a revolution that has been taken most seriously, by navies worldwide, is the one in Military Affairs or RMA. Of course, whether there is, in fact a revolution underway or not has also been a subject of debate. It has been suggested that military technology has actually been evolving over centuries, and examples of the crossbow, the horseman's saddle and stirrup, gunpowder, tank, wireless, aeroplane and missiles are given as an example of this evolutionary continuum. So what is all the excitement about?

If we accept that "revolutions" are marked by non-linearity in the progression of events, then it has to be accepted that military affairs are indeed in the throes of a revolution.

Adapted from the Keynote address delivered at a seminar on Transformational Technologies for the Future Navy, organized by the Republic of Singapore Navy during IMDEX 2005, Singapore, May 17, 2005.

Since about the mid-1980s it has been apparent to discerning military observers that galloping advances in weapon, sensor, platform and information technology are indeed taking us on an exponential path. Having come to terms with RMA, militaries have started to internalize it, and that is how the term transformation has recently entered their lexicon.

For navies, transformation encompasses not just the creation of capabilities by the use of technology, but also changes in organisational relationships, war fighting concepts and peacetime doctrines or standard operating procedures. Transformation is essentially based on a willingness to constantly challenge old thinking and introduce new concepts. This obviously means that the focus cannot be exclusively on technology and hardware, but must also embrace people.

Access to technology and the ability to absorb and exploit it will have a vital impact on the effectiveness of a Navy. A note of caution needs to be sounded at this juncture.

Transformation, as it is being talked about today, presumes the ready availability of cutting edge technology, and substantial funding as well as infrastructure to support the introduction of new concepts into service. With a few exceptions, the vast majority of navies are not only faced with a shortage of funds but also lack access to advanced technologies.

In this context, however there are two points to note. Firstly, we must remember that it is up to us to take as much or as little of transformation as we need. And secondly, we all have our strengths as well as weaknesses, but the synergy that results from cooperative engagement often renders disproportionate results. The need of our times is for neighbouring navies to work together to ensure maritime security for common benefit. Hence, in my opinion, while transformation is looked at, by individual navies they should also examine it in a wider frame of reference and see whether a possibility exists for collective transformation in which many of us can participate.

It will be my endeavour here to gently interweave cooperation with transformation. We have common maritime concerns, which should lead us to cooperate while attempting to transform our forces, and operations.

Before touching upon some of the transformational technologies required for tomorrow's navies, it would be worthwhile examining some emerging strategic trends in the maritime arena, which I would term as drivers for maritime cooperation.

Firstly, globalization is cutting across national boundaries and leading to the evolution of linkages between cultures, economies and peoples. This growing economic integration is evident from increasing trade interdependence, and measures like the decision to establish an ASEAN Economic Community by 2020. Consequently, to borrow John Donne's words, no nation remains an island anymore "entire of itself" - except perhaps in a purely geographical sense. Collective and not individual prosperity is the way of the future.

Secondly, the critical importance of energy security, which implies continuous and assured supply of energy resources such as oil, natural gas and coal, cannot be underestimated. Any challenge to the free flow of energy can lead to major conflict, which will have profound effects on regional and global economies. Ensuring energy security is therefore, a major maritime issue of common concern.

Thirdly, low intensity maritime threats, which include piracy, gunrunning, drug smuggling, illegal immigration, etc., are increasing in intensity and frequency. Being transnational in character, these threats require the cooperative use of maritime forces to tackle them effectively. India already conducts joint patrols with Indonesia and Sri Lanka and an agreement with Thailand for similar patrols is due to be signed soon. The danger of a grave threat from the seas is also driving security initiatives like the International Ship and Port Facility Security (ISPS) Code, the Proliferation Security Initiative (PSI) and the Container Security Initiative (CSI). These initiatives require multilateral maritime cooperation in order to succeed.

And lastly, environmental issues like the sustainable use of fishing resources, and pollution control, as well as matters relating to Search and Rescue, EEZ Patrols, etc., are increasingly engaging the attention of navies and coast guards. All these issues also need a multilateral approach to be effective.

Against the backdrop of what I have said so far, oceans today are not expanses of water that divide continents, but maritime highways that link nations. I also see them as a broad canvas on which the drama of international relations is often played out.

Our other adversary, of course, can be Mother Nature, and we always need to be prepared for natural disasters. It is just five months since this region went through the tragedy of the Indian Ocean Tsunami. This disaster afforded an opportunity to demonstrate clearly to the world community the unique global brotherhood of the seas, and the ability of navies to not just work alongside each other, but also to catalyze and facilitate cooperation amongst nations.

I want to point out that in contrast to the Cold War era, when it was ideology that created hostility, the fight today is against the forces of anarchy, obscurantism and fundamentalism. If there can be an unholy nexus of international proportions amongst the bad guys; men of goodwill too must come together. Navies, therefore, must be equipped with the wherewithal to enable international cooperation. Let us then have a look at a few of these technologies and start with a survey of naval platforms in all the three dimensions.

First of all, the surface ship. This platform has weathered many debates about its continued utility in the face of emerging threats of every nature, including growing costs. Experience has, however, shown that large multipurpose ships allow great flexibility across a range of activity extending from hot conflict to humanitarian aid. Warships also provide unique reach and sustainability. Keeping these useful attributes in mind, technologies relating to propulsion, low visibility and hull forms must engage our attention.

One of the lessons of the 2004 Tsunami was that navies must have the ability to access the coast easily from the sea, which is not always permitted by existing hulls. Developments have been undertaken to find unconventional hull forms that would combine shallow draft, higher speeds, better sea-keeping qualities, and improved survivability. The new ideas that find application in warships include: multiple hulls like catamarans, trimarans and pentamarans, surface effect ships, air-cushion vehicles, semi-planning monocoques, and delta hull forms, etc. Stealth is another consideration that is literally changing the "shape of ships" to come.

The displacement of the battleship by the aircraft carrier was one of the most significant maritime developments of the last century. This is one platform that has remained at the vortex of controversy, and continuously attracted fire from its many detractors. However, quite apart from their military utility, events have shown that aircraft carrying platforms - and I include the LHD, LHA and LPD -have a tremendous role to play in peacetime activities too, especially disaster relief.

The Indian Navy has now been operating carriers, for close to half a century, and we have just reaffirmed our faith in this platform by buying one from Russia, and commencing the construction of a second one in India. Carriers have a whole range of unique technologies associated with aircraft launch, recovery and navigation, which are also evolving rapidly.

Factors such as high acquisition, manning and lifecycle costs of aircraft have led navies to look towards unmanned aerial vehicles (UAV) to supplement the efforts of Fleet Aviation. The Indian Navy was amongst the first to have operated UAVs at sea, and we have found them most useful. The future seems to point towards an increasing role for such unmanned platforms - in the air, underwater and on the surface - to perform cooperative surveillance tasks with greater efficiency and reduced costs.

Due to the submarine's inherent stealth qualities; sea denial and trade warfare have remained its forte. Earlier attempts to integrate submarines with fleet operations were largely stymied by communication problems, but with increasing connectivity and longer range weapons, submarines are increasing their scope of operations. Even as interest in submarines grows rapidly amongst navies in the IOR, technology is being marshalled to bring about major improvements in the performance of the diesel- propelled boat.

Areas of focus include noise reduction, non-hull penetrating and optronic masts, towed wire antennae and sonars, cruise missiles, launch of Special Forces, and of course air independent propulsion. Technology will ensure that the submarine continues to hold its position as a platform of the future; and transformation will provide new and more useful roles for it.

The search for new means of propulsion at sea is driven as much by the rising cost, as by the environmental impact of our dwindling fossil fuel resources. Nuclear propulsion has its uses, but only in limited applications. With advances in materials such as permanent magnets and high temperature superconductors, the focus has now shifted to electric propulsion and many propulsion plants are already at sea, but mainly in the cruise liner business. The progress of electric propulsion is being watched with great interest, and the future certainly seems to lie in the "all electric ship".

With modularity rapidly gaining acceptance, both as an economy measure and a force multiplier, platforms will become multi-role. Such designs would allow swapping of guns, weapon launchers, mine counter-measures gear or helicopter decks as required. Modular systems like the Danish STANFLEX are an idea whose time has at last come; and navies are being increasingly drawn towards it, as is obvious from the Littoral Combat Ship and "Streetfighter" under development for the US Navy.

The missile, torpedo and even the lowly mine continue to be an omnipresent threat to the naval combatant. Ship and submarine-launched cruise missiles, as well as shore based anti-ship missiles add to the complexity of maritime operations in the littoral. As smart weapon systems enter service, even smarter counter-measures are on offer, but budgetary constraints may not always permit navies to invest in them. Therefore, tactical acumen has to often substitute for technology at sea.

The vast benefits of multimedia telecommunications have made it possible for fleets to have enhanced knowledge management. Not only is the operational domain becoming increasingly transparent but the need for distant units to share a common situational awareness amongst themselves and with the command ashore is driving the need for networking.

Networking a navy is an undertaking as fraught with complexity as it is expensive in execution. The Indian Navy has however, taken the plunge and committed itself to networked operations before the end of the decade. In the context of regional maritime cooperation, there is no area more urgent than the establishment of communication channels and protocols that enable better coordination. Information Technology offers us a convenient means of linking friendly forces, and we also have a satellite launch capability in the region. Therefore, in the years ahead, we could perhaps look at a regional maritime data network. This would enable navies to cooperate effectively in areas such as SAR, pollution control, countering maritime terrorism and piracy, and of course, disaster relief operations.

Finally, warship building today is an expensive proposition due to the high infrastructure costs of shipyards, as well as the expense of design and development. Collaborative R&D, design, building and even marketing efforts between two or more countries in this region can bring with it many advantages, including economies of scale, amortization of development cost and simplification of logistic support.

Conclusion

As one contemplates the mechanics and implications of transformation at sea, it is easy to lose oneself in the marvels of technology. In all this, there is a need to retain clarity and focus, and to bear in mind that technology is only a means to an end.

Moreover, while exploring technology and exploiting it at sea, we must also try and leverage it to bring navies together. The sum of the parts is often greater than the whole, and with the tremendous resources, and human talent available, given the resurgent economies of our region, we can achieve a lot if we work together.

As we go about our business at sea, those of us in our respective navies must work assiduously to create synergies for mutual benefit. If the unfortunate tsunami episode had a bright side at all, it was to provide us a heartwarming glimpse into the extended family system of South and South-East Asia, where countries reach out in times of adversity.

In conclusion, I would like to reiterate that the imperatives of globalization and our growing dependence on sea resources will ensure that the 21st century is a Maritime Century. Navies, if they so choose, can become catalysts for peace, tranquility and stability in the Indian Ocean Region. In this endeavour, transformational technologies will be their handmaidens.

Ski-jump Operations at Sea
A Great Leap Forward

O n July 18, 1976 in compliance with the Montreaux Convention of 1936, the Soviet Government informed Turkey of the imminent southward passage of *Bolshoi Protivolodochny Kreyser* or large anti-submarine warfare (ASW) cruiser from the Black Sea. However, Western observers noted with a great deal of excitement that the Russians were in breach of the convention because the ship that actually entered the Strait of Bosphorous was the Red Fleet's first aircraft carrier, the *Kiev*.

This 45,000-ton ship was, at that juncture, the largest warship ever built in the USSR and carried 15 to 20 each of Yak-36 vertical take-off and landing (VTOL) fighters, and Ka-25 helicopters. In addition, it had a massive and varied array of weapon systems, sensors and a comprehensive electronic warfare suite. This kind of a fit left no doubt that the ship was intended to operate as a self-contained, long-range unit in a high threat environment.

Threats Prevail over Dogma

Actual capabilities apart, the significance of the *Kiev* lay in the fact that the Soviets had, at long last, felt constrained to relent in their dogmatic opposition to aircraft carriers, and to deploy integral air power at sea. This could have not been possible but for the availability of VTOL fighters to the Soviets. The *Kiev* was not really viable if pitted against a US attack carrier, but once the concept and the technology were proven, it was expected that improved versions of ships and aircraft would be forthcoming. And they were.

The Soviet Navy's role till after World War II had been purely defensive and subordinated to the Red Army's need for protection of its seaward flank. It, therefore,

Air Power, (The Journal of Air Power and Space Studies), Volume I, Number I, Monsoon 2004, pp. 1-11.

saw no need for ship-borne air power, and was content to operate under the cover of its powerful shore-based naval aviation, the Morskaya Aviatsia, in waters contiguous to Soviet territory. US carriers were not only ignored but also denigrated as "sitting ducks" for Soviet missiles. Accordingly the Soviet naval strategy over the 1950s and the 1960s was centred on anti-carrier measures like missile-armed long-range bombers and cruise missile-equipped ships and submarines. But a ship-borne air strike capability did not figure in their perceptions at this juncture.

In the mid-1960s, the induction of the Polaris-armed nuclear submarines gave a new dimension to the threat perceived by the USSR. The threat was partially countered by the build-up of the long-range anti-submarine aircraft like the IL-38, TU-95 and TU-142 in its air arm. It also appears to have been the catalyst for initiating a change in Soviet thinking about the desirability of putting air power on seagoing platforms.

The first manifestation of this change of heart was the commissioning of the ASW helicopter carriers *Moskva*, and *Leningrad* in 1967-1968. Around 1972, US satellite reconnaissance pictures began to show a large carrier type ship under construction in the Black Sea Shipyard of Nikolayev. Closer scrutiny indicated that she was not a carrier in the traditional sense. Absence of launch and arrester gear showed that she was intended to operate helicopters and possibly VTOL aircraft. This was the *Kiev*, commissioned in early 1976, followed by the *Minsk, Baku* (renamed *Gorshkov*) and *Ulyanoy*. Of these ships, the only survivor now is the Gorshkov, which has recently been acquired by the Indian Navy and is undergoing modernization in the port of Severdovinsk on the White Sea. A logical continuation of the carrier-building programme was the Orel class. Originally planned as attack carriers with nuclear propulsion, only one ship of this class was commissioned in 1991 and remains in Russian Navy service. A second ship, the *Varyag*, was terminated half way through construction and sold to China.

Soviet Aviation Goes to Sea

Towards end-1989, Western reconnaissance agencies observed unusual maritime activity in the Black Sea. A Soviet aircraft carrier of much larger dimension than the Kiev class was undertaking trials involving what appeared to be conventional (as opposed to VTOL) fixed-wing flying operations. Incredible as it sounded to many, reports indicated that the aircraft participating in these trials were fourth generation, shore-based supersonic fighters like the MiG-29 and Su-27! These reports proved authentic, and it is known that this was a new 67,000-ton carrier, the *Tiblisi* (later renamed *Kuznetsov*) where this new concept of operations was proved. What appeared to be a "rags to riches" story bears further examination here.

During the latter part of the Cold War, Soviet Navy destroyers had been known to closely "mark" from very close ranges, United States Navy (USN) and Royal Navy (RN) aircraft carriers during North Atlantic Treaty Organisation (NATO) exercises, causing immense annoyance and anxiety to their commanding officers. They did this in order to observe (and to film) aircraft launch and recovery operations for days on end. However, this was not enough to master the esoteric art of carrier aviation, which they had scoffed at for over half a century It, therefore, must have become obvious to the Soviets at some stage, that in order to bridge the tremendous expertise gap that existed, they would have to leapfrog the era of conventional take-off and landing (CTOL) aircraft and adopt VTOL technology in order to bring their aviation to sea. And this they proceeded to do through the flat deck *Kiev* class carriers and the Yak-36/38 VTOL aircraft.

The second combat jet VTOL aircraft in the world to enter service (after the Harrier), the Yak-36 was not entirely a satisfactory solution to the problem of fleet air defence at sea. Unlike the Harrier, which made use of a single engine with vectoring nozzles for forward flight as well as hovering and landing, the Yak-36 had one large engine for forward propulsion, in addition to two smaller lift engines which were used for VTO and shut down thereafter. This three-engine configuration not only added to the basic weight of the aircraft and limited its range/endurance but also increased the demands on piloting skills during take-off and landing.

Like the Harrier, the Yak-36 too lacked the engine thrust and aerodynamics to attain supersonic speeds. The Harrier, because of its swiveling nozzles, could, however, perform a short take-off (STO) from deck, which greatly enhanced its load-lifting capacity as well as range/endurance. The Yak-36, due to its engine configuration, on the other hand could take off and land only vertically. Thus, the Harrier was classified as a V/STOL aircraft, whereas the Yak-36 remained just a VTOL machine.

It essence, though they had a proved concept, the Soviet Navy realized that the Kiev/Yak-36 combination would never be a match for a USN carrier battle group deploying immensely capable fighters like the F-14 Tomcat and F/A-18 Hornet supported by E-2 Hawkeyes and A-6 Prowlers. There was need to put more capable combat aircraft on their carriers.

The Russian Approach to Technology

Those familiar with weapon platforms of Soviet/Russian origin know that the Russians are highly innovative in their thinking. Frequently circumscribed by their own technological limitations, they are known to produce systems, which are quite unconventional and imaginative. Their solutions to design problems are sometimes crude but functional, and often based on a "brute force" approach.

Nothing illustrates this better than the early version of the MiG-29, which had a conventional (i.e., stable) aerodynamic configuration, and was equipped with a set of hydraulic controls, but could more than match the instant turn rate and agility of its (aerodynamically unstable) digital fly-by-wire Western contemporaries like the F-16 and Mirage-2000. This was due as much to good aerodynamic features, as to the massive thrust of its power plant, which could overcome the build up of induced drag at high angles of attack.

The Soviet design bureaus had produced conventional fourth generation combat aircraft, many of them superior to their Western counterparts. Their naval architects and shipyards now had a proven capability for delivering aircraft carriers. Yet the question that confronted the naval staff was how to adopt a 20-30 ton combat aircraft, originally designed as a land-based fighter, for operations from an aircraft carrier.

It is well known that a ship-borne fighter can operate with ease from a shore base. On the other hand, a combat aircraft designed to take off and land on 10,000 feet of concrete runway cannot possibly operate without major modifications from the 700-800-foot-long flight deck of a carrier. The main problem areas in such an undertaking would be excessive landing and take-off speeds, inability of undercarriage and airframe to withstand the stress of carrier recovery, and finally, dimensional incompatibility to fit the ship's aircraft lifts and hangar.

Above all, to enable aircraft operations, the ship would need to be equipped with a hydraulic arrester gear for landing, and a steam driven catapult for assisted take-off. Neither of these complex systems had ever been designed or built in the USSR at that juncture.

Russian Solutions

These were problems of a magnitude that may have been daunted Western expertise, but the Soviets apparently took them in their stride. Three types of aircraft, already in frontline service with the Soviet Air Force, were selected for conversion to ship-borne operations: the MiG-29 and Su-27 fighter/attack aircraft and the Su-25 ground attack aircraft (which saw extensive operations in Afghanistan). Significant aerodynamic design changes were undertaken to reduce take-off and landing speed. The undercarriage was strengthened for deck operations, and an arrester hook fitted in a reinforced fuselage underbelly. Where necessary, wing and tail folding mechanisms were incorporated to reduce dimensions and enable stowage on board.

The aircraft designations were suffixed with "K" or "G" for "Korabelnyy" and "Gak"; the first standing for "ship-borne and the second for "hook". On November 1, 1989 the MiG-29K, and the Su-25G, all successfully landed for trials on the new carrier *Kuznetsov*, using a Soviet designed arrester gear.

The Ski-jump Concept

All these were absorbing developments for military analysts, but what really took their breath away was the fact that the three aircraft had been launched from the ship, not with the help of a catapult, but over a 12 degree ski-jump, integral to the carrier's deck. The ski-jump concept, in the Western doctrine, was firmly wedded to V/STOL operations, and while some experimental flying had possibly been done, little thought had been given to using it for routine launch of conventional aircraft from a carrier.

The ski-jump was the brainchild of a young Royal Navy engineer named Lieutenant Commander Doug Taylor, who hit upon this idea in the early 1970s while investigating ways of enhancing the short take-off and payload performance of the Harrier in order to enable deck operations from small carriers. He faced considerable cynicism from aircraft designers as well as test pilots, because the simple device appeared to promise "something for nothing" and everyone felt that there had to be a catch somewhere.

In the mid-1960s, amidst a major national debate, the Labour government imposed drastic defence cuts, which saw most of Britain's aircraft carriers decommissioned, and naval aircraft transferred to the Royal Air Force (RAF). Interestingly, a number of RN admirals had resigned on principle, on the emotive issue of the virtual disbandment of their fleet air arm.

A decade later, the RN, desperate to bring about a revival of its air arm, saw the *Invincible* class aircraft carriers and the Harrier V/STOL fighter as a heaven sent opportunity for its resurrection. The *Invincible* class ships were actually a ploy by the RN to reintroduce aircraft carriers by subterfuge. The project had originally been projected to the British government in the guise of "through" (or continuous) deck "cruisers" designed for operating anti-submarine helicopters to meet NATO tasks. In actual fact, they were nothing but small aircraft carriers. At 20,000 tons, however, they could not possibly operate Phantoms, Buccaneers or any of the other contemporary carrier aircraft then available. Moreover, these ships were powered by gas turbines and, thus, could also not be equipped with a steam catapult for launching aircraft.

However, with the advent of the Harrier, there emerged a distinct possibility of putting fighter aviation back at sea. The problem was that these ships could provide only a very short deck run for launch of a fighter. So they decided to investigate Taylor's ski-jump concept seriously.

A shore-based ski-jump with a hydraulically variable profile was fabricated and installed at the Royal Aircraft Establishment, Farnborough (subsequently shifted to the RN Air Station, Yeovilton) and a series of trials proved that Taylor's idea did indeed have substance. The Harrier's performance from a marginal small deck carrier was to see dramatic enhancement in payload, and reduction in take-off run by using this device.

This ski-jump has been thereafter retained in Yeovilton to train fledgling pilots (which included the author, 20 years ago) for ship-borne operations.

Basically, the ski-jump consists of a curved inclined ramp installed in the bows of the earner. An aircraft traversing the ski-jump follows its curved profile, and on exit, this trajectory (for the same take-off run) can place the aircraft 200-300 feet higher in the air and, thus, provide it more height (and time) to accelerate into forward flight. However, because of the short take-off run available on a ship's deck, an aircraft would normally exit from the ski-jump at a very low speed. Should this speed be below the stalling speed of the aircraft (as it is for the Harrier), the flying controls remain ineffective, and the aircraft will, in layman's terms, "fall out of the sky."

In this respect, V/STOL aircraft have an advantage, because at low speeds, they rely not on aerodynamic controls (which require high relative air flow over the aerodynamic surfaces), but on jet reaction controls. These controls, called "puffers", use hot air, bled from the engine, and allow the machine to be controlled in all planes till it accelerates into an aerodynamically safe flying regime. It was for this reason that use of the ski-jump could not be contemplated for CTOL aircraft, under normal circumstances.

Some of the most grateful beneficiaries of the ski-jump were perhaps Indian naval aviators. We had learnt to fly the Sea Harrier from the 12-degree ski-jump of HMS *Hermes* with bitterly cold winds blowing in the English Channel. On return to India, we looked with horror at the flat deck of INS *Vikrant* (with a catapult at the end). On a normal (hot) day, even with a partial fuel load, full deck run and water injection, the aircraft could accelerate to barely 85-90 knots, and exit the deck at just 50 feet above the waves. It made the launch quite exciting, but stressful for pilots and left no margin for any errors-especially at night. Subsequently, when the ship was fitted out with a 12-degree ski-jump, we could launch with full load from half the deck run, and the aircraft would effortlessly reach a height of about 250 feet at the end of the launch.

Marrying CTOL Machines with the Ski-jump

Conventional wisdom (in the West) dictated at this point of time that CTOL aircraft could be launched from a ship's deck only with the help of a catapult, which would accelerate it to a respectable speed of 120-140 knots within a run of 150-200 feet. One of the more complex and trouble-prone pieces of seagoing machinery, this device has evolved over five decades through hydraulic, pneumatic and explosive versions into the present day steam powered catapult which is currently manufactured only in the USA. Given their ingenuity, there is no doubt that the Soviets would have eventually come up with their own (perhaps electrically driven) version of an aircraft catapult. However, they obviously had other views on the subject.

Soviet origin aircraft have suffered by comparison with their Western contemporaries in terms of aerodynamic finish and sophistication. However, as pointed out earlier, these lacunae are more than made up by the massive thrust that pours out of their powerful (albeit smoky and fuel guzzling) aero-engines. Thus, aircraft like the MiG-29 and Su-27 have a "thrust to weight" ratio of better than unity, which results in a short take-off run and very rapid acceleration once airborne. In all likelihood, it is this attribute, combined with life augmentation devices and a digital "fly-by wire" system, which has made these CTOL aircraft controllable at sub-stall speeds and encouraged the Russians to contemplate a ski-jump launch.

Since they had decided to move on from performance limited VTOL aircraft like the Yak-36, the Soviets devised an ingenuous solution, which consisted of bypassing the catapult route, and using the ski-jump for launch of CTOL aircraft. For recovery of these aircraft, they resorted to the tried and tested method of using hydraulic arrester gear consisting of wires stretched across the deck. This operating mode gave rise to a new acronym in the naval aviation lexicon: STOBAR, which stands for short take-off but arrested recovery.

In order to test this concept, as well as the aircraft modified for punishing carrier operations, the Soviets established a comprehensive test facility at Saki, near Sevastopol in Crimea. Partly through emulation and partly through their own innovation, the Soviet Navy installed a ski-jump and an elaborate hydraulic arrester wire system, complete with their own versions of visual aids and an electro-optical landing sight. It was here that the STOBAR concept as well as the Su-27K, MiG-29K and the Su-25G were proven before they started deck trials on the *Kuznetsov*.

During a visit to Saki some years ago, I had the opportunity to fly off the Russian ski-jump and do an "arrested" landing. An inspection of the elaborate facility created by the Russians for training carrier pilots was an eye opener, and provided a most interesting comparison with the approach and concepts used by the British, nearly two decades earlier.

The Future of STOBAR

The Russian Navy has been in dire financial straits for some time now, and its commitment to carrier aviation must remain a question mark amidst all its other preoccupations. After the extensive deck flying trials programme undertaken by the *Kuznetsov* in 1993-1994, the ship was equipped with an Air Group consisting of Su-27K fighters and Ka-28 helicopter. She sailed into the Adriatic during the Bosnian crisis, but has not been seen very often at sea thereafter. With the ship tied up, it must be difficult to keep the crew, especially the pilots, in a reasonably operational state. The carrier training facility in Saki (now in Ukraine) could be a big help in this respect, provided the Ukranians permit its regular use.

Three of the Kiev class aircraft carriers have been disposed off, and the last of the Orel/Kuznetsov class, the *Varyag*, valued at US$2.4 billion was sold to Chinese Macao, ostensibly as a "floating casino". The Russian Navy is possibly at a crossroads now and it does not appear likely that the country has either the funds or the political will to launch any major shipbuilding projects. But no matter what happens in the future, the past two decades would have been an exciting and fulfilling period in the brief lifetime of its fixed-wing carrier aviation arm. Two specific achievements can be counted as landmarks in aviation history and stand witness to the ingenuity and innovation of Soviet/Russian designers and engineers: the rapid conversion of shore-based CTOL combat aircraft into equally capable carrier-borne versions, and the iconoclastic demonstration that CTOL aircraft can be safely operated from ski-jumps carriers.

As far as the Indian Navy is concerned, the die has been cast. We were amongst the first to use a ski-jump operationally, and have, over the past 20 years, built up great faith in the concept. The former *Gorshkov,* during her modernization will be equipped with a ski-jump and a set of arrester wires to enable STOBAR operation. The MiG-29K that our pilots fly off her deck will have little resemblance to its Indian Air Force (IAF) ancestor of similar designation. Apart from a tail-hook, this fourth generation aircraft will have digital fly-by-wire controls, a glass cockpit, modern multi-function radar, a refueling probe and much-enhanced range/endurance. A carrier-borne version of the light combat aircraft, designated LCA (Navy) is under development at Aeronautical Development Agency (ADA) Bangalore, and will also operate in the STOBAR mode as a strike/fighter, first from the *Gorshkov* and eventually from the indigenous aircraft carrier being built at the Cochin Shipyard.

While the US Navy, with its super-carriers and wealth of Naval Air Systems (NAVAIR) capabilities, has watched the rise and decline of Russian Naval Aviation with a degree of detachment, its poorer cousin, the British Royal Navy must have taken keen interest in these developments. The future British aircraft carrier [designated CV (F)], on the drawing board at the moment, is projected to be a 40,000-50,000-ton ship and selection of an aircraft has been a vexed issue for some time now. Among the options being actively discussed for the $ 5 billion CV (F) is that of STOBAR operations. The aircraft that the CV (F) operates may well be a version of the joint strike fighter, which can be launched from ski-jump, and land into a set of arrester wires.

If imitation is the best form of flattery, the Russians should certainly have reason to be pleased.

India's Quest for An Indigenous Aircraft Carrier

It was indeed fortuitous for the Indian Navy (IN) that at the moment of India's independence, those charged with planning for the nation's embryonic maritime force included many men of vision. In 1948, within six months of freedom, a ten-year naval expansion plan had been prepared, largely by Royal Navy (RN) officers serving on secondment, for consideration by the Government of India.

The plan was drawn up around the concept of two fleets; one for the Bay of Bengal and the second for the Arabian Sea, each to be built around a light fleet carrier, to be replaced subsequently by a fleet carrier. The grandiose scheme provided for four fleet carriers and 280 ship-borne strike and fighter aircraft in the next few years. This plan received approval in principle from the Governor General Lord Mountbatten, as well as the Prime Minister Jawaharlal Nehru, but unfortunately failed to materialize because of a variety of reasons.

Hostilities with Pakistan in Jammu and Kashmir in the winter of 1947 had focused India's attention on the Himalayas rather than the oceans and the young nation's scarce resources were being diverted to the army. Moreover, the outbreak of the Korean War in 1950 required all World War II naval aviation surpluses of the RN to be marshalled once again for service, and there was not enough to spare for the IN. However, the main impediment that emerged was a change of heart in His Majesty's Government.

The reason why the British were willing to help bolster India's naval strength was the basic assumption that the IN would form part of a Commonwealth-based bulwark against any possible depredations by the Soviet "Bear" in the maritime domain (perhaps looking for warm water ports?). However, by now India had decided to adopt an equidistant non-aligned stance between the two "cold-warriors" and her steadfast refusal to be part of any military alliance considerably dampened the RN enthusiasm and support.

RUSI Defence System Journal, July 2006, Vol. 9, No. 1.

Arrival of the First Carrier

The dire financial straits of the fledgling nation also posed many hindrances to the planning process for creating a navy, but the carrier project eventually survived, albeit in a drastically diluted form. In 1957, the unfinished hull of HMS *Hercules,* a Majestic Class light fleet carrier was acquired by India and taken in hand for completion in Belfast. Four years later she was commissioned into the IN as INS *Vikrant.* Sister ships served in the Canadian and Australian Navies for some years.

The aircraft complement of *Vikrant* was essentially a Hobson's choice, and consisted of the Armstrong-Whitworth (later Hawker-Siddeley) Sea Hawk, a straight-wing, first generation ground-attack jet fighter, and the French built Breguet Alize turbo-prop ASW machine. Both the aircraft served us faithfully, but by the late 1970s they were long in the tooth and needed urgent replacement.

The Royal Navy's dilemma arising from the Defence White Paper of 1966, which saw her naval aviation being virtually dismantled, was mirrored to an extent in India. The RN had been deprived of both fixed wing carriers and aircraft; the IN had a small carrier but could not find a new aircraft that was compatible.

The one single factor that led to the resurrection of the fleet air arm of the RN was the, "discovery" of the ski-jump by a young engineer named Lieutenant Commander Doug Taylor. It was only with the ski-jump installed on board that the small "through deck cruisers" of the Invincible class could become viable STOVL (Short Take-off Vertical Landing) platforms. We joyously followed suit, and the first IN Sea Harrier FRS Mk. 51 was vertically landed on the deck of *Vikrant,* off Goa by the author in December 1983.

In the hot and light wind conditions of the Indian Ocean, undertaking Short Take-off operations from the flat deck of *Vikrant* in a Sea Harrier (with one outrigger on the catapult track) always remained an "exciting" event, especially at night. There was much relief in the aircrew-room therefore, when she was fitted out with a 12 degree ski-jump by Mumbai Dockyard in 1985. The range and payload of the Sea Harrier received a much needed boost too.

Search for a Second Carrier

With the *Vikrant* getting on in years, there was serious concern in the IN that having constructed our operational and tactical doctrines around carrier aviation, we may, one day find ourselves without a fixed wing platform at sea. Apart from the operational penalties, what our planners dreaded most was the inevitable loss of flying and other expertise, painstakingly built up over three decades of shipboard aviation at sea. Carriers were obviously not going to be available off the shelf and a serious thought process was triggered off in Naval HQ regarding the design and construction of an indigenous ship to meet our long-term needs.

The immediate scenario however, remained bleak till 1985, when we received an offer from the RN, for the sale of the 28,000 ton HMS *Hermes*. Laid down in WW II, this 25-year-old ship had served the RN as a fixed wing, commando and eventually STOVL carrier. The offer was eagerly accepted, and after a refit in Devonport, the Falklands flagship (wrongly claimed to have been struck by an Argentinean Exocet) sailed for India as INS *Viraat*. She had a 12 degree ski-jump and was well adapted for STOVL flying operations. Some of us had already flown from her deck during the Sea Harrier conversion in Yeovilton.

Vikrant was finally de-commissioned in 1997, and the IN has since been operating with just the *Viraat* at sea, with her complement of Sea Harriers FRS MK.51, Sea King Mk. 42 ASW helicopters and the Indian built version of the Allouette III known as Chetak. Occasional visitors on board include the Kamov-28 (ASW) and the Kamov-31 (AEW) helicopters.

Indigenous Endeavours

The Indian Navy's small Directorate of Naval Design was in the 1970s, deeply engrossed in the exciting maiden venture of licensed production of the *Leander* class frigates being undertaken by the Mazagon Dockyard, Mumbai. From time to time, they did however, toss about various aircraft carrier design options and a concept for the conversion of a passenger ship hull to a "flat-top" was briefly examined in 1979-1980, but discarded since it received no encouragement from any quarter.

By 1987, the IN had persuaded the government to approve the commissioning of a concept study by *Direction des Construction Nauale* (DCN), France of a sea control ship of about 25,000 tons, capable of operating aircraft in the 15 ton category. The DCN report received in 1989 covered two options; one of a conventional (catapult equipped) ship and the other of a ski-jump carrier, to be constructed at the Ministry of Shipping owned Cochin Shipyard. The report unfortunately came in at a time of financial stringency, and had to be reluctantly shelved by the IN.

However, the DCN exercise was not entirely futile because it gave a fillip and inspiration to our own designers, and concept designs, first of a simple 16,500 ton "Harrier-Carrier", and then of a larger, more versatile 20,000 ton ship emerged from the Directorate. The factors and choices for size/configuration of the ship formed a rather complex matrix, and a little digression is needed to dwell on them.

The Options Available to India

Apart from the dimensions of the hangar, the size of the propulsion plant, and capacity of fuel tanks and magazines, the most important determinant of carrier, design is the flight deck, whose size and configuration depend on type of aircraft operations intended.

John F. Lehman, former US Secretary of the Navy, in his book, *Aircraft Carriers: the Real Choices,* written in 1978, provides some very useful empirical data. Studies have shown that to operate all conventional high performance aircraft a deck length of 912 feet is required, and this would correspond to a displacement of about 60,000 tons. If heavier aircraft like the F-14 were excluded, the deck length could be reduced to 813 feet with the ship displacing about 35,000 to 40,000 tons. Lower down the scale, a 650 to 700 foot deck would suffice for purely STOVL operations and the ship would displace about 20,000 tons.

In the 1980s and 1990s, the choices of aircraft available to India were severely circumscribed on account of political considerations. Carrier-borne aircraft of US origin, by far the most capable in the market, were then just not available to us. The Soviets, our main purveyors of military hardware at that time, had only one shipboard fighter — the three-engined VTOL fighter Yak-36 (Forger) to offer, but experience showed that the Sea Harrier, already in our inventory, was superior in most aspects. And then there were two more options, both at different stages of development: the French Rafale-M shipboard fighter and the Indian designed Light Combat Aircraft (LCA).

However, one factor emerged clearly; aircraft catapults were manufactured only in the USA and since this piece of machinery was unlikely to be made available to India, we could discard ship designs, which were based on conventional aircraft requiring a catapult launch. This eliminated all US origin deck aircraft as well as the Rafale as viable options. Since we had already decided that the Yak-36 did not have much merit, our ship designers were placed in limbo once again. This is when the ingenuity of the Russians came to our rescue.

Enter STOBAR

Towards the end-1980s, word started trickling out of the USSR of an unusual experiment being undertaken by the Morskaya Aviatsia, the air arm of the Soviet Navy. Having shed their dogmatic animus of many years to flat-tops, the Soviets were planning to make a dramatic entry into the arcane field of carrier aviation. They planned to shun the trodden path and do this through the medium of ski-jump equipped carriers. But what about the flying machines?

Having realized the limitations of their VTOL endeavours, they chose three (conventional) shore-based combat aircraft and undertook extensive modifications to enable ramp take-offs and hook-assisted "arrests" on board. The aircraft chosen were the Sukhoi-25 (trainer and strike aircraft), the Sukhoi-27 and the MiG-29. The modified versions of the aircraft were given the suffix "K" (for *Korabelnyy* or ship) and this mode of operation added a new term to the lexicon of naval aviation: STOBAR, which stood for "short take-off but arrested landing".

The Air Defence Ship

With this development, our carrier design options began to acquire some clarity, and the Staff Requirements having been reviewed, the designers returned to the drawing board. However, the continuing uncertainty about aircraft availability, made their job difficult, and the first tentative design that emerged was for a 20,000 ton carrier named euphemistically the "Air Defence Ship" or ADS, (partly to camouflage it from air force snipers). The ADS would operate the Sea Harrier (already in our inventory) and hopefully the indigenous LCA whose ship-borne version was being explored.

However, a detailed feasibility study of a STOBAR version of the LCA by its design bureau revealed that a safe ski-jump launch and arrested recovery, though feasible, would make extra demands on this radical little strike-fighter. Although equipped with a digital flight control system, the delta wing configuration of the LCA (Navy) would require higher take off and landing speeds. Consequently the deck length had to be increased by about 15 meters, and the redesigned ship now displaced 24,000 tons, with a corresponding increase in cost.

By now the IN was seriously examining the Russians offer of their 1980s vintage helicopter/VTOL carrier *Admiral Gorshkov,* and a choice had to be made of a suitable aircraft. The obvious options were the Su-33 (a derivative of the Su-27K selected for operation from the 67,500 ton carrier, *Kuznetsov),* and the MiG-29K. An evaluation revealed that both aircraft would meet our operational requirements. The Su-33, though more capable, being dimensionally larger would not only not fit in the smaller hangar of the 44,500 ton *Gorshkov,* but would have marginal wing-tip clearances from the island structure during deck launch. It was, therefore, decided that the MiG-29K would equip the *Gorshkov,* to be renamed INS *Vikramaditya,* once it entered Indian service.

The downstream impact of this decision was instantly felt by the ADS programme, and a fresh design exercise was initiated to assess the implications of MiG-29K STOBAR operations on the ADS design. According to the planners, the ship's basic complement would be a squadron each of MiG-29s and assorted helicopters. The option of operating the upgraded Sea Harriers was also catered for, till the LCA (Navy) received its full operational clearance. The workshops, magazines, deck and lift configurations as well as crew spaces had to be re-worked.

The staff requirements having been finalized in 1999, the ADS emerged, in its definitive form, as a 37,000 ton vessel, to be powered by four LM-2500 gas turbines in COGAG arrangement which would give it a top speed of 28 knots. The 830 foot long angled flight deck would have a set of three arrester wires aft rated to handle aircraft of up to 22 ton all-up weight. A set of jet blast deflectors and hydraulic chocks would be installed to provide a 600 foot deck run for launch of the MiG-29K and LCA (Navy), from the 14 degree ski-jump launch using afterburner. The ship would carry an air group of 30 aircraft and helicopters and would be crewed by about 1,400 personnel.

The Indigenous Aircraft Carrier

The project having received financial approval of the Government of India in January 2003, the first steel was ceremonially cut in Cochin Shipyard on April 11, 2005. At this time the ADS was re-designated as the "IAC" or Indigenous Aircraft Carrier. Consultancy for propulsion system integration will come from M/S Fincantieri of Italy (now in the final stages of completing the Italian carrier *Count Cavour)*, and for the aviation complex from M/S Nevskoie Design Bureau of Russia.

Some early problems relating to ship-building steel and selection of equipment have been resolved and the yard is optimistic about meeting the delivery schedule of 2012-2013. There are no illusions that this is going to be a complex undertaking, and on account of certain residual imponderables the shipyard plans to execute the contract in two phases. It is expected that the uncertainties, especially those relating to equipment that needs to be imported would have been resolved by the time work starts on Phase II. The financial estimates for the IAC have, therefore, remained somewhat flexible so far.

As a practicing adherent of ship-borne aviation for the past 45 years, the IN aims to fulfill its long-term operational commitments in the IOR by deploying two carrier task forces at sea, while a third ship is under maintenance or refit. This would be the embodiment of a concept mooted in our plans as far back as 1948.

The arrival of the *Vikramaditya* and her squadron of MiG-29K fighters in 2008 would certainly add considerable combat power to the IN, and the Service looks forward to the IAC joining the fleet in the next decade. However, while we take up a case for construction of a second IAC, we would need to assess the residual life of the sturdy old *Viraat.*

Building an aircraft carrier for the first time is no doubt going to be a challenging task for India's warship designers and builders. The commissioning of this ship in the next decade would not only be a defining event for our industry but also a concrete manifestation of the determination and resolve with which we have pursued the vision of becoming a "Builders Navy".

WARSHIP BUILDING IN INDIA
A Reappraisal

It is often overlooked that India's maritime tradition goes back to a few millennia before the Christian era. Our west coast has historically seen intense trade and commercial activity being undertaken by sea, with the Persian Gulf, Mediterranean and the East African littoral. From the East coast successive dynasties which ruled peninsular India up to the 12th century AD, sent waves of adventurous seafarers to spread Indian culture and civilization to South-East Asia, where it is very much in evidence even to this day.

A maritime tradition can only survive on a sound ship-building industry, and here we need to remind ourselves that we are the proud inheritors of the world's oldest dry-dock built during the Harappan period (2,400 BC) in Lothal, Gujarat. While the ancient dhow-building tradition of our West coast ensured that Indian vessels were ubiquitous in the Indian Ocean, many generations of the Wadia family of master shipbuilders sent Bombay-built sloops, schooners, merchantmen and men o' war sailing the seven seas.

Today, India's economic resurgence is directly linked to her dependence on trade and commerce, most of which is conducted by sea. It is vital, not just for India's security but also for her continued prosperity, that we posses a Navy which will protect the nation's vast and varied maritime interests. The Navy's role is to help maintain peace in the Indian Ocean, meet the expectations of our friends and neighbours in times of need, and underpin India's status as a regional power.

Whenever an Indian-built warship sails into a foreign port today, it receives looks of admiration, not unmixed with surprise that a Third World industry is capable of such sophistication. We are fortunate that the seeds of a self-reliant blue-water Navy were laid by our farsighted predecessors when they embarked on the brave venture of undertaking modern

---

Adapted from the Keynote address delivered at a seminar on Warship Building, organised by the Garden Reach Shipbuilders & Engineers, New Delhi, March 22, 2006.

warship construction in this country four decades ago. We certainly need to acknowledge that our shipyards have done us proud by delivering 85 warships to the navy in this period.

The first *Leander* class frigate was built under licence in Mazagon Docks in 1972. The basic hull form has thereafter been stretched, broadened, re-designed and re-armed by our ingenuous naval architects, and eleven ships later, we have seen the progressive metamorphosis of INS *Nilgiri* into INS *Beas* commissioned in 2005. Armed with a hybrid and eclectic weapon suite, this frigate is arguably one of the most unique and powerful warships in her class today.

Currently, the Indian Navy has on order, 27 ships which include fast attack craft, landing ships (tank), frigates, destroyers, submarines and an aircraft carrier; and there are more in the pipeline. In addition, the Indian Coast Guard has its own acquisition plans. I doubt if the shipbuilding industry of any other country can look forward to such an attractive and mouth-watering" prospect.

It is an article of faith in the Naval Headquarters that we will create a "home grown" navy, and our commitment to indigenization in the long term, is total. We are today offering a unique opportunity to the country's ship-building industry, along with its ancillaries, to help us build a great Navy commensurate with India's stature.

At the same time, it is to be stated clearly that we are talking about maintaining the Navy's force levels, which is an issue that impinges on national security. Without being alarmist, I would like to sound a note of warning that unless we can accelerate our warship production; we may be heading for a crisis. The bottom line is that the Navy's force levels have to be maintained; and if our yards are unable to deliver quality warships on time, we will have no choice but with great reluctance to buy them from abroad.

In this context, let me flag two or three issues, which have great relevance:

Firstly, the endemic delays and cost overruns, which have dogged our warship building programmes are beginning to take their toll. While the ultimate cost of our warships is still cheaper than the rest of the world, the actual sums involved are rising rapidly in successive projects. Those who sanction our budget are becoming sceptical, and we are soon going to have a crisis of confidence on our hands, unless we can bring far more accuracy to our forecast of project costs and times. Both the Navy and the industry need to reflect deeply on this issue, because I know from my personal experience that on one hand:

- People who draw up Staff Requirements in Service HQs suffer from two syndromes. They have the habit of asking for the "best" when the "good enough" would do equally well.

- They also have a mortal dread of obsolescence creeping up on the equipment that they have selected. So they simply refuse to freeze the Qualitative Requirements till the programme starts to slip badly.

- On the other hand, the shipyards on their part use such delays, not only to cover up their own slippages, but also add to the costs by billing the project for an idle workforce and infrastructure.

Secondly, as far as the functioning of our ship-building industry is concerned, their work ethic, efficiency and productivity remain rooted in the past. It is possible that full order books tend to generate complacency and that is perhaps why one does not often see signs of commercial, financial or technical innovation. 1 feel that much more can be done by the industry, as far as introduction of modem technical practices, as well as better financial, and human resource management are concerned.

I would strongly urge the Department of Defence Production and the Ministry of Shipping, to unshackle our shipyards from bureaucratic constraints and to empower their management with as much freedom of action as possible.

At the same time, 1 would also exhort the Managing Directors themselves to be bold and innovative in their approach. You have the opportunity of a lifetime to leave your imprint on a national undertaking. Try to emulate the Indian industry outside which is performing miracles and attaining world-class standards. Your work force fully understands the reality that they must enhance productivity or perish; it is for you to go to them and enlist their full support.

The last point I wish to discuss is the concept of "self-reliance". We know that colossal resources in terms of skills, expertise, and infrastructure that need to be mustered in order to produce a modern warship or submarine, it is obvious that the undertaking is no longer going to be an activity that can be confined to a single shipyard. The self-sufficiency that we seek, should therefore, be in certain well-defined core areas, and in our quest for self-sufficiency we should not waste time and resources in re-inventing the wheel. Whenever required for better efficiency, and wherever it will save time, we should enlist outside expertise or consultancy from India or abroad.

With current order books full, and an ambitious warship-building programme in prospect, our shipyards need to evolve solutions to deliver ships in much shorter time frames. There is obviously, an urgent need to explore new strategies, in terms of partnerships with other enterprises within India or even globally in order to achieve better efficiency in production and management.

Our private sector has many strengths and has proved itself in the international arena. The time has now come to invite the private sector to contribute to warship building by creating public-private partnerships, and joint-ventures, or even leasing, outsourcing and off-loading as required. The powerful synergy of both public and private enterprise must be harnessed for national good.

Geo-politics and Foreign Relations

A Vision for the Andaman & Nicobar Islands

I was last here in early January 2005, when the destruction wrought by the monstrous earthquake and tsunami loomed large over these beautiful islands that we know and love so well. Flying over the entire chain then was akin to seeing a ship that had taken a torpedo hit amidships and was not only sitting low in the water, but also listing. Except that a tsunami carries the punch of a tactical nuclear weapon rather than a torpedo.

Car Nicobar Island was of course the 'Ground Zero' of the island chain and visiting it revealed nature's raw power at its strongest and devastating worst. But even nature, powerful as it is, cannot conquer the indomitable human spirit. And here while paying solemn tribute to the memory of the gallant IAF personnel who lost their lives, one must acknowledge the courage of those who survived and stood fast in the face of danger. Today, in the best tradition of our Armed Forces, IAF Station Car Nicobar is back in business, and is fully operational.

That is why it has been so satisfying to revisit the island yesterday and see how much has been done to repair the damage wrought by the earthquake and tsunami. The rebirth and rehabilitation of settlements also reveals the spirit and vitality of the people of these islands, who have shown tremendous courage and fortitude in the face of adversity, and the resilience to start life once again from scratch. Against the backdrop of this huge tragedy we watched from the mainland with admiration, as the Andaman & Nicobar Command (ANC) stood firm as a rock and rendered yeoman service to our stricken fellow citizens with promptness, efficiency and compassion. But for your valiant, sustained and untiring efforts, in support of the civil administration, not only would the rescue and relief efforts have taken longer, but also the rehabilitation and return to normalcy would have perhaps been delayed.

Adapted from the Keynote address delivered at the seminar, 'A&N Islands: Vision 2025', Port Blair, August 25, 2005.

Now that the islands have emerged from the shadow of fear and destruction, it is appropriate for us to look ahead and plan a better future for the islands and their inhabitants.

Forward planning, of course, has never been our forte as a nation. And indeed, when we do plan, implementation remains an equally weak area. These islands represent a microcosm of India and their small size and population makes it relatively easy to implement a plan for their development, provided, of course, that there is transparency and honesty of purpose in its implementation. Indeed, if a vision for these islands could be encapsulated in one phrase, it would be to develop them as "Islands of Excellence", in every sense. The scale of operations being small, it is much easier to prove concepts here. So if an experiment succeeds in the Andaman & Nicobar Islands (A&N), we can establish islands of excellence elsewhere in the country too.

India's vision for the effective use and development of these islands has two major aspects to it. First of course is the issue of their internal economic development, so as to make the islands financially self-sustaining and bring prosperity to the islanders. The second, and equally important issue is their vital role in the national security matrix, because of their strategic location.

Not many people on the mainland seem to be aware of the fact is that the A&N islands constitute the southern and eastern frontiers of the Indian union. Closer to South-East Asia than peninsular India, their unique location in the Bay of Bengal throws up challenges as well as opportunities. They could be lucrative objects of desire for any country, which may harbour ambitions of dominating the Bay of Bengal. At the same time, they form a springboard from where India can reach out and project power, exert influence or strike bonds of friendship in our eastern neighbourhood. In this role, the Armed Forces, through the ANC, have a vital role to play.

Vision for Economic Development

To identify the potential for economic development of any place, one should first identify, what is now termed in management jargon as its "USP" or unique selling proposition. The USP of these islands is their pristine maritime environment, their rich maritime resources, their eventful past, a uniquely valuable anthropological heritage, and of course, their strategic location vis-à-vis major shipping routes

As you are aware, the A&N Islands have a long and interesting history. Situated as they are, at the crossroads of major trading routes, mariners over the centuries have invariably stopped by and left their mark. The first recorded reference to the islands was in the 2nd century AD by the Roman geographer Ptolemy, who named them the islands of "good fortune". The Chinese monk, I-Tsing in the 7th century and Arabian merchants in the 9th century also spoke of the inhabitants as fierce and cannibalistic. The eunuch Chinese Admiral Cheng Ho is said to have visited the islands in the 15th century on one of his voyages, with his famous armada and in the 17th century the islands witnessed a brief spell of Maratha rule. Though not hospitable, these islands often served as a watering station or refuge from the elements, for European seafarers in the 17th and 18th centuries. Ultimately the British, in search for a penal colony assimilated them into their Indian Empire.

In the Indian psyche, the A&N will always be associated with the dark symbolism of "Kaala Pani" as this penal colony came to be known. This is where the political opponents of British Raj were sent in the company of hardened criminals, often never to return home. Just four months after Pearl Harbour, the Japanese invaded and occupied the islands as part of their quest for a "Greater Asia Co-prosperity Sphere". In fact, very few people are aware that these islands were the first part of India to be liberated from British rule when Netaji Subhas Chandra Bose visited Port Blair and hoisted the tricolour on December 30, 1943. Of course, this was a temporary reprieve because the British reoccupied the Islands after the Japanese surrender in 1945. This chequered past, lends these islands an aura of fascination which is a draw for tourists.

Even more important than the contemporary history of these islands is their anthropological wealth. The Andamanese Negritos, the Onges, the Jarawas and the Sentinelese constitute a living time capsule of the evolution of human beings, which is not available anywhere else in the world today. Few in numbers, barely understood, difficult to approach and on the verge of extinction, the Andamanese tribes are among the most unusual people alive today and their number is steadily decreasing. No other living human population has experienced such long-lasting isolation from contact with other groups, an isolation that has lasted, on genetic evidence, from between 30,000 to as much as 60,000 years. It is worth remembering that 30,000 years ago the world was still in the grip of the last ice age, mammoths roamed in Eurasia and the Neanderthal man was still living. It is highly probable that the Negritos represent the least changed survivors of the first migration of homo sapiens into Asia. Theirs is the most ancient surviving way of life on earth today.

Their uniqueness can be gauged from the fact that in all, there are less than 500 Andamanese Negrito today while the population of India has crossed one billion. The Andamanese Negrito represents a period of Indian pre-history so ancient that it makes even the earliest Indian epics look recent. The Negritos in general and the Andamanese Negrito in particular represent major genetic evidence in the scientific search of the origins of modern humanity. They are thus, not just the "senior citizens" of the world, anthropologically speaking, but also constitute a precious heritage, which belongs to all mankind.

Greater attention, therefore, needs to be paid to the preservation of these tribes without disturbing their natural way of life, for they add immeasurably to the existing diversity of our nation and lend mystique and charm to these islands. The soul-stirring image of a Sentineli warrior attempting to spear a Coast Guard Helicopter, which had gone there to check on their welfare after the tsunami, epitomises the essence of human spirit as embodied in these noble savages, who remain self-confident, self-sufficient and unapproachable. This spirit cannot be allowed to die and must form an integral part of the vision for these islands. After all, first and foremost, these were their islands.

Coming to the wealth of natural resources that lie under the waters of these islands. The EEZ around the A&N comprises 30 per cent or one third of our total economic zone, and just for comparison, is more than the EEZ of Pakistan. There is said to be considerable potential for finding hydrocarbons here, and evidence is already available, of gas nodules in the seabed, which can be profitably mined for natural gas. This means that economic development should be built around the exploitation of the islands' natural resources, as well as the use of the pristine maritime environment for tourism.

Another USP of these islands is their strategic location in close proximity of the major shipping routes. This leads many of us to contemplate their use as a trans-shipment hub, a bunkering facility, or a duty free port, with visions of replicating Singapore in mind. Such visions may not be a practicable proposition, because the one thing that these islands definitely lack is a hinterland. We also need to remember that since the pristine environment of these islands is a major asset, any development along these lines will bring with it, major penalties in terms of environmental degradation, and may not be acceptable nationally or internationally.

With the limited scope for industrial and agricultural activity in the Islands, tourism remains the thrust sector for revenue and employment generation in the islands. Keeping in view the fragile ecology and limited capacity of the islands, the vision must be to develop the A&N Islands as an up-market island destination through environmentally sustainable development of infrastructure, without disturbing the natural eco-system, with the objective of catalyzing socio-economic development of the islands.

There are many ways in which this can be achieved. We could open more sanctuaries and National Parks for promoting eco-tourism or nature tourism. Yachting and ocean sailing are booming as a sport activity. We definitely need to look at state-of-the art marinas, with facilities for yachtsmen, along with developing a terminal for cruise liners at Port Blair. Construction of eco-friendly semi permanent jetties at popular destinations combined with the development of adventure water sports complexes could be another thrust area.

The concerns of environmentalists will have to be sufficiently addressed by identifying, sequestering as many areas as necessary for the preservation of bio-diversity, in the form of sanctuaries and nature parks. There is also need to firmly enforce environmental laws and engender a respect for nature amongst our citizens. Thereafter, at some point of time, the Supreme Court will have to be approached to relax the draconian measures imposed by the judiciary in an effort to protect the bio-diversity of the A&N.

The waters of the islands abound in marine life, which in addition to an abundant harvest of edible fish also contains many rare and valuable species. As you all know, this bounty of the seas attracts poachers from far and wide in this region, and requires a massive effort on the part of the Indian Navy and Coast Guard to police and protect. One of the main reasons that foreign trawlers are attracted to these waters is the lack of an organised indigenous fishing industry. Public or private investment in motorized trawlers, cold storages, and canning factories on one or more islands could not only change the face of the local economy, but also keep poachers at bay. Poachers as you are all aware, pose a problem not only at sea, but also become a law and order problem in Port Blair once released from jail.

I am sure, that with careful and proper planning, the islands can not only become self-sustaining, but also generate a surplus of funds, which could pay for the development of modern infrastructure in the form of widespread telecommunications, all-weather harbours, good roads, etc.

While it is prudent to have a vision for the future, it is equally important to focus on the needs of the day. We need to introduce incentives for harvesting of rain water, recycling of waste water, solid and liquid waste management systems, and other eco-friendly practices. Incentives also need to be given for promoting the use of non-conventional energy resources, and environment friendly building guidelines.

Having created a vision for the economic development of these islands, let us turn our attention to the place of these islands have in India's strategic calculus.

Strategic Vision

The best way of visualizing the strategic importance of these islands is to imagine our security situation if they were not a part of India, or indeed if they were to be taken over by a hostile power! And this is not as far-fetched as it may seem. About two years ago, there was discussion in the Pakistani press that at Independence, the A&N islands too should have been divided between the successor states, with East Pakistan getting its due share. This kind of talk may appear ridiculous, but should in actual fact give us cause for thought and perhaps concern. So let me first highlight a few factors of geography, which are often overlooked by most Indians.

In the north, the islands are separated from Myanmarese territory by a mere 22 nm. At the southern end, Indira Point lies just 90 nm from the troubled Aceh province of Indonesia. Mainland Myanmar is just 170 nm away and Thailand only 270 nm. In stark contrast to this proximity, Port Blair is over 750 nm from the Indian peninsula.

Of the 573 islands, islets and rocks in the chain, only 38 are inhabited. The north to south spread of these islands extending over seven degrees of latitude enables them to command the Bay of Bengal. Of particular significance is the fact that the six degree channel at the western end of the Malacca Strait is dominated by the Great Nicobar Island. Although the islands extend over an area of only 8,300 sq km, their geographic location confers on India, an additional EEZ of 300,000 sq km.

One should envision these islands as a bridgehead to South-East Asia and beyond. This role is well illustrated by the use of the islands for building bilateral defence relations in the recent past. The MILAN exercises instituted in the mid-1990s were a perfect complement to our "Look East' policy and resulted in sending a powerful positive signal to our maritime neighbours, of our peaceful intentions. The escort of US high value ships in the Malacca Straits in 2001-2002 also played a significant role in bringing India and the US closer together.

At the moment, the Bay of Bengal seems placid enough, and anyone raising the bogey of a security threat can be termed alarmist. However, if there is one lesson that the events of the past 50 years should have taught us, it is that we cannot take national security for granted. Perceptions of national interest are difficult to assess and keep changing with circumstances. Therefore, the next best thing is to keep a careful watch on others' capabilities, and to keep one's own powder dry. Let us have a quick look at the strategic scenario in the region.

Coming first to Myanmar. A bilateral maritime boundary agreement between India and Myanmar in 1987, and a trilateral agreement involving India, Thailand and Myanmar in 1993, saw the resolution of disputes over the Cocos Islands, which went to Myanmar, and Narcondam Island, which remained with India. Myanmar is an important neighbour with whom we share land as well as maritime boundaries, and it is to our mutual benefit that the two countries maintain close and cordial bilateral relations. Actions of nations are guided by their own vital interests and it is not for India to say whom Myanmar should choose as her friends or allies. However, it would cause us considerable concern if any outside power were to find its way into the Bay of Bengal and the Andaman Sea through Myanmar, since this is an area of vital interest to us.

It is against this background that we need to strengthen our relations with Myanmar, and to build mutual confidence. The Myanmarese have given clear indications that there has been some rethinking and that they would welcome a closer defence relationship with India. Therefore, while the Ministry of External Affairs (MEA) works out its plan of action, the Armed Forces can certainly assist by accelerating the pace of defence cooperation. The Navy has already taken up a number of proposals, and we are confident that things are moving on the right lines. As it happens, the ANC is right up in front and should form the spearhead of our friendly overtures to Myanmar.

Relations with Thailand have blossomed over the past few years with the signing of a bilateral trade agreement and initiatives like BIMSTEC. In July 2005, we signed a Navy-Navy MoU for Coordinated Patrols along the maritime boundary in the Andaman Sea, which is due to commence in September 2005. The Royal Thai Navy (RTN) is also keen to get our assistance in keeping their small force of naval Harriers going and we hope that this will be the first step in building a closer bilateral relationship. Once again, the ANC will have much to do with the RTN, especially with their 3rd Naval Area in the days to come, and we hope that we will be able to build a meaningful and enduring relationship between Port Blair and Phuket for a start.

After a period of indifferent relations, economic reasons have been the catalyst for the revival of Indo-Malaysian ties. We have had a long-standing Defence MoU with them, but this remained at a low-key until last year, when it was revived; and things have been looking up since. My visit to Kuala Lumpur in July 2005 set the stage for Staff Talks between both Navies as well as closer cooperation between the Air Forces of both countries. We are hopeful that Indo-Malaysian relations will evolve with rapidity.

Our largest maritime neighbour and the one who is less than a hundred miles from the A&N Islands is Indonesia. Here again, after years of neglect, bilateral defence ties took an upswing with the signing of a Defence Cooperation Agreement between the two countries in 2001. Consequent to this Agreement, our ships have undertaken hydrographic surveys in Indonesian waters. Fears of insurgency in the troubled province of Aceh spilling over, also prompted both countries to commence Coordinated Naval Patrols in contiguous waters, which are continuing since the past two years. Bilateral relations are warm and during my visit to Jakarta, a number of additional steps have been agreed upon, which include possible sale of defence equipment, steps to establish intelligence sharing over terrorist movements and increasing the scope of bilateral patrols.

If to all this we add our very strong economic and defence ties with Singapore, it is clearly evident that relations with our South-East Asian neighbours are better than they have ever been. What we need to guard against now, is the serious menace posed by non-state actors, gunrunners, poachers, illegal immigrants and the like. The use of these islands as sanctuaries or trans-shipment bases for drugs smuggling and gunrunning by terrorist organisations is also a distinct possibility. Illegal immigration in particular is an area of concern because the immigrants can become a fertile ground for anti-national activity.

We would not like another Kargil-like situation to ever emerge in these islands, and prudence demands that we remain prepared for all eventualities. However, the sensitivities of our neighbours have to be kept in mind while considering any accretion of forces in the ANC. Hence, as far as military preparedness for security of the islands is concerned, rather than actually positioning assets here, it is more important to develop the infrastructure to be able to support them effectively, when the need arises. Therefore in the planning process, we need to look ahead and ensure that we create the basing, maintenance, support, and logistics facilities for an appropriate army, navy and IAF force levels. In the long term, planners should take into account a division sized formation, one and a half fighter squadrons, and major warships including a carrier and submarines in transit.

By the very nature of these island territories, and keeping in view the rising threat of low intensity conflict at sea, what actually needs to be enhanced is our reconnaissance capability, coupled with better intelligence assets so that we can keep the air and sea space around these islands under surveillance on a continuous basis. We also need sufficient airlift and sealift capabilities and means of rapidly accessing the islands through the development of jetties, airstrips and helipads.

While developing security assets, we must constantly bear in mind the ever present threat from the vagaries of nature. The precious lessons learnt from the tsunami experience must be embedded in our operational and logistics planning processes. The Indian Navy has used the experience to evolve SOPs, and published a Book of Reference for disaster relief. The episode has also impacted on our force planning process. For one, the Government has readily accorded approval to our proposal for building or acquiring a small number of LPDs. These ships will have a capacity of about a thousand troops, and will carry heavy-lift helicopters for ship-shore movement, and are thus ideal vessels for the dual tasks of amphibious operations as well as disaster relief.

ANC as a Crucible for Jointness

The invaluable role being played by the ANC as the crucible for the concept of jointmanship is something I would like to touch upon. From a 150 man naval garrison in 1962, to Fortress A&N in 1976, and then through the Government directive of October 2001, to a full-fledged Joint Command, this formation has travelled a long way in four decades. Along with it, the Indian Armed Forces too have taken a great leap of faith when they placed all forces located in the A&N Islands, including the Coast Guard, under the command of the newly created, C-IN-C AN, who reported to the Chairman COSC. The roles and functions allotted to the command included:

- Defence of the territorial integrity, waters and airspace of the islands.

- Ensuring that eastern approaches to the Indian Ocean remain free from threats for unhindered passage of shipping.

- Monitoring of SLOCs in designated AOR.

- Exercising surveillance over EEZ.

- Establishment of an ADIZ for air defence and air space control.

- Undertaking joint planning for contingencies and infrastructure planning.

The unwritten charter of the ANC, which I consider even more important, was a conceptual one. It was to evolve, from first principles, a framework for a unified formation and to test it in the crucible of the command along with working rules. This framework, if successful, would then become the template for replication elsewhere, especially if theatre commands were contemplated.

ANC is technically under the Chairman COSC, but committees are not known for speed of decision-making, especially where policy matters are concerned. Presently, operational matters and routine issues relating to budget, works and personnel are referred to the HQ IDS, which gives its full attention to this command, and we have an efficient system in place. At the same time, a linkage is also maintained with the parent Service HQs through the unique Service Component Commanders, for manpower requirements and maintenance of assets.

There is no doubt that ANC has been an unqualified success as an experiment, and has proved beyond doubt that the concept of Jointmanship can work very successfully in the Indian environment. If conclusive proof of this was necessary, it was provided during the tsunami relief operations when the well-oiled and efficient machinery of the ANC came to the assistance of the civil administration, and the people of these islands. The seed of jointmanship planted in Port Blair four years ago is now a sturdy young sapling, whose shoots can be transplanted anywhere else.

Whether this happens or not, is a moot question. Realization came to armed forces across the world many years ago that for reasons of economy, efficiency, and combat-synergy no single Service would be able to undertake operations on its own in the future. Most military powers have followed in the wake of the USA to some extent or the other. However, each country has its own compulsions, and a different environment to manage. Perhaps we are not yet quite ready for jointmanship in India, but it can only be a matter of time. And when that time comes, it will be the ANC which will be the cradle, the school, and the torch-bearer of jointmanship for our armed forces.

Conclusion

In a slow moving and ponderous democracy like ours, nothing is either straightforward or simple. Add to this, the distance of these islands from the seat of power, the lack of intra-island communications, the high cost of any undertaking in the A&N, the major environmental concerns, and the verdicts that come from our pro-active judiciary; and the problem appears insurmountable. But we must not lose heart. We must look at how other countries manage such contradictions to strike a balance between the environment, national security concerns and the fundamental good of our people. That sums up the challenge we face.

Indo-US Relations
A View from Seawards

I return to the hallowed precincts of this great institution, a decade and a half after leaving it. I note with pleasure that the "big blue bedroom" has retained its familiar colour and relaxing ambience. I have no doubt that it continues to exert its hypnotic charm and soporific influence on the students. My return has been nostalgic and although I spent only ten months in Newport, it seems almost like a homecoming. We made many enduring friendships, internationally in the Naval Command Course. That's what this course is all about. But let me add that the warmth, friendship, kindness and hospitality that my wife and I received from Americans have stayed with us till today.

One has to keep reminding oneself of the harsh reality that the world has changed tremendously in the last 15 years. I recall that there was a bunch of us in the Class of 1990, who made it a point to be photographed on top of the highest building in every city we visited during the tours of Continental USA. There is one photograph of the 6 or 7 of us on the roof of the World Trade Centre. That this building no longer exists is a tragic reminder of our changed world.

The world of *circa* 1989-1990 that we lived in, studied about and analysed, in the College seminars was a very different world from that of today. However, dramatic changes were taking place even then. In early 1989, the Soviet forces had completed their withdrawal from Afghanistan. In November that year, the Berlin wall collapsed. In February 1990, the Soviet Communist Party agreed to give up power, and the Commonwealth of Independent States was just months away.

In the War College Review of Autumn 1990, the President Rear Admiral Strasser had commented on the Iraqi invasion of Kuwait and said, "Regardless of the outcome of the

---

Adapted from a talk delivered at the Naval War College, Newport, Rhode Island, USA, March 24, 2005. The author did the Naval Command Course at Newport in 1990.

current crisis, the world will never be the same again." Earlier, a young Director in the US State Department named Francis Fukuyama, in a now famous article, had declared that mankind faced "the end of history as such", because he felt that mankind had evolved to a stage where Western liberal democracy had become the "final form of human government".

Time has moved on, and the world is indeed a different place today, but history has certainly not ended, nor is Western liberal democracy anywhere near being the universal norm. However, I am not here to provide you a worldview, but to acquaint you a little more closely with my country, and specifically touch upon the maritime relationship between India and the USA. The best and perhaps kindest word to describe Indo-US relations over the past 60 years would be "oscillatory". They have often swung from warm and cordial to cold and frosty, which is a pity, because our two countries have so much in common. But international relations have their own logic, and I am no expert.

India's Past

To most foreigners, India seems to present many contradictions and conundrums. I could spend a long time trying to explain all of them. But to obtain an understanding of present day India, it is necessary to delve a little into her past. And I will not go too far back.

India attained her Independence from two centuries of British rule in 1947, and today, she is 58 years old as a modern, sovereign republic. However, as a cultural, religious and civilizational entity with an unbroken history, she is perhaps a hundred times as old. Being a colony, we had missed the industrial revolution completely, and at independence, the founding fathers of this populous and poor nation were faced with many dilemmas and difficult choices. One of these was; which model of economy should be chosen to run the fledgling nation?

Our first Prime Minister Jawaharlal Nehru was a lawyer, educated and trained in the UK, and a Fabian at heart. Under his leadership, it was decided that in order to ensure social and economic justice for India's poor masses, we would follow, what was termed a "Socialistic pattern of society". This essentially meant a western style liberal democracy, but with state control of the levers of economy and industry. This model also strived to attain early scientific and industrial self-reliance by import substitution.

We therefore became, for the first four decades of our existence, a closed and somewhat stagnant economy. While a great deal of progress did take place in many fields like agriculture, health and education, we realized that entering the global economic mainstream could have given us a much more accelerated growth. In 1990, an acute financial crisis forced the Government to take some radical steps, and while gingerly opening up the economy to globalization, started the process of privatizing state owned enterprises. Today, India's economy is forecast to become the fourth largest in the world by 2020, and the third largest by 2050.

Along with our erstwhile Socialistic economy, another area, which created a great deal of confusion in the West, was India's policy of "non-alignment". In the sharply polarized world that existed in the middle of the last century, this was merely a manifestation of our desire to remain equidistant from the two power blocs that existed. We were certainly not a Communist state, and at the same time, as a newly independent colony we were equally reluctant to join our former imperial masters and become a member of the many treaty organisations, which bound the Western world. However, many in the West saw our failure to jump on the bandwagon as a sign of bad faith. We ourselves did not.

Yet another area, which intrigued Western observers and led to considerable confusion, was our use of Soviet hardware and weapon platforms. A Tupolev patrol aircraft or a Kashin class destroyer always had a connotation of "enemy" unit about it, so when these were seen in Indian colors, there grew a perception in the Armed Forces of the West that we were part of the Soviet bloc. Some went so far as to say that we had Soviet "advisers" working with us, or even had leased bases to them.

Nothing could have been further from the truth, and our arms transfer relationship with the Soviets had its own genesis. India's post-independence leadership was a little utopian in its outlook, and considered expenditure on defence to be wasteful. In 1962, a border dispute with China led to a short war in the Himalayan terrain of India's north. Our troops fought bravely, but compared to the experienced PLA, they were ill equipped and came off second best in the conflict. This setback brought home the realization to the Indian politicians, that for far too long they had ignored national security and starved the Armed Forces of funds and modern equipment. We went looking for modem fighters, tanks, ships and submarines, and the first countries we approached for these arms were the UK and the USA.

The response that we received was less than enthusiastic. Many in the West at that juncture saw things only in black and white. The concept of non-alignment did not appeal to them, because they were of the view that "if you are not *with* us, then you are obviously *against* us". So the Indian request for arms was met with a very lukewarm offer of World War II surplus equipment, which we politely turned down.

In stark contrast, the Soviets offered us top of the line weapons at very reasonable rates and liberal credit. We also sent our personnel for training to the USSR whenever we received a new system. Our relationship with the USSR remained purely economic, and to an extent military, but it did give us a "pink" tinge in Western eyes.

In the past two decades we have made determined efforts to diversify our sources of military hardware, and succeeded to a great extent. We have also embarked on a campaign for indigenisation on long-term basis. Since these measures will take time, in the foreseeable future, the Indian Armed forces will continue to have a mix of Russian, Western and of course, Indian equipment.

India's Maritime Dimension

With that brief introduction to India's recent past, let me now turn to the country's maritime dimension. India is the only country to have an ocean named after it, and a brief examination of our rich maritime history provides an insight into the central role that the Indian Ocean has played in shaping the country's destiny. Let me illustrate this by a quote from K.M. Panikkar, who in a book entitled *India and the Indian Ocean*, written in 1945, says, "...millennia before Columbus sailed the Atlantic, and Magellan crossed the Pacific, the Indian Ocean was a thoroughfare for commercial and cultural traffic. The close connection between the early civilizations of Nineveh and Babylon, and the West coast of India, and the analysis of ancient artifacts found in the Indus valley demonstrates the extensive use of the sea in the period around 3,300 BC."

In ancient India, while the West coast witnessed mainly commercial activity, the Eastern waters were used as a medium for establishing more enduring relationships with South-East Asia, but not by invasion or conquest. The deep cultural linkages of the subcontinent that are still in evidence with Myanmar, the Malayan peninsula, the Indonesian islands of Java, Sumatra, and Bali, and with Thailand, Laos and Cambodia stand witness to this.

The plains of north India are rich and fertile, and have historically invited foreign invasions. A majority of the invasions and migrations to India came overland through the mountain passes. First came the Aryans, to be followed by Greeks, Parthians, Sythians, Huns, Turks, and early Christians. But such incursions either led to transient political changes, or to the foundation of new dynasties, which in time were absorbed by the cultural fabric of India and became Indian. On the other hand, the invaders who came across our shores by sea were the ones who stayed to rule, to exploit and then to return home with our wealth.

It was the decline of India's sea power by around the 14th century, in the Mughal era that was largely responsible for the success of European adventurers who began to arrive on our shores in the next century. The Portuguese came first, followed by the Dutch, British and the French. It is, therefore, etched deep into the psyche of the average Indian that the final domination of the country by an alien power resulted not from overland invasion, but by incursions over her shores. India did not lose her independence till she lost the command of the sea. This is possibly a factor in India's post-independence insularity and attempts to attain autarchy.

The British did maintain a naval force in Indian waters, which through various incarnations has become today's Indian Navy. But in sharp contrast to a powerful Indian Army, which was trained, nurtured, and deployed by them worldwide, they were very careful not to allow this Navy to grow beyond a limited coastal force.

At Independence in 1947, this force was further depleted when it had to be split up between the two successor states: India and Pakistan. We started off with a very basic brown water navy of a handful of destroyers, frigates and minesweepers. But we were fortunate to have a visionary leadership, which despite all odds ensured the development of a compact but balanced three-dimensional navy. Within a decade we had acquired frigates and an aircraft carrier from the UK, followed by submarines and other ships from the USSR, and a couple of tankers from Germany. By the early-1970s we had the elements of a modest blue water force in place.

Two salient developments on the maritime front can be flagged at this juncture. Firstly, in the 1960s we decided to embark upon a warship-building programme. Starting with the British Leander class frigates, Indian yards have till today produced 73 warships and submarines of imported and indigenous design.

Secondly, the country's political leadership began to realize the potential of sea power as an instrument of state policy, and to make use of it - albeit somewhat tentatively. The Indian Navy's role in the liberation of Bangladesh in 1971, bolstered the country's confidence considerably, and manifestations of this were seen in subsequent deployments to help maintain Sri Lanka's integrity in 1987, and to restore the legitimate government of Maldives in 1988.

The Indian Navy's rise coincided with the end of the Cold War, which had hampered India's strategic options because of bi-polarity. The post-Cold War period has seen a coordinated use of the Navy in concert with certain foreign policy initiatives such as intensifying engagement with the US Navy, or a "Look East" policy which aims at enhancing our relations with countries in the Asia-Pacific like Myanmar, Singapore, Vietnam, Indonesia and Japan.

What I have attempted to bring out by this account of India's maritime past is, the key role of maritime power in our relations with the Asia-Pacific region. The underlying historical thesis is that when India was strong at sea, the resultant commercial and cultural interaction benefited everyone in the region. However, when external forces gained a stranglehold over our seas, the result was a domino-like fall of states in the region to colonization. With the re-emergence of the Indian Navy as a maritime force of consequence, there is a growing recognition of India's critical role as a pillar of security and stability in this vital region.

India does not subscribe to spheres of influence, or regional doctrines, but history, geography, demographics and economics do have their own compulsions. India's current national objectives lie in ensuring a secure and stable regional environment free of tensions, which will permit the economic and social development of our masses, and the fulfillment of her destiny. If this requires insulation from external intervention, then the Navy must be prepared to act as the outermost security cordon.

India's Maritime Environment

Let me now point out to India's maritime interests, which have their own security implications. The country has a strategic geographic location in a warm water ocean, and sits astride major shipping routes from the Persian Gulf to the Malacca Straits. A 7,500 km long coastline studded with over 200 major and minor ports and a rich hinterland, a well endowed EEZ of over 2.1 million sq km, and as many as 1,200 islands off both seaboards are some of our maritime assets, which can become liabilities in times of tension. Further afield, we have interests in the Antarctica where we have established research stations since the early 1980s.

Additionally, reforms in the past decade have opened up India's economy, which is now integrated with, and interdependent on other world economies. Our trade has been growing at a very rapid rate, and is forecast to double every five years or less. Almost 80 per cent of this trade by value, and over 90 per cent by volume is carried by sea. Trade accounts for only 8.5 per cent of our GDP at the moment, and the tremendous scope for expansion can be gauged by the fact that it currently forms less than 1 per cent share of world trade. A significant portion of India's trade is carried by about 600 Indian merchant ships whose rapidly growing tonnage adds up to over eight million tons and is expected to hit ten million shortly.

Over 70 per cent of India's crude oil is imported; all of it by sea. Our consumption is expected to rise 150 million tons per annum by 2040, making India the largest importer of oil in the world. Quite apart from our own interest in the oceans, the world's economic lifeline runs past India's doorstep. Over 60,000 ships are known to transit through the Indian Ocean every year, transporting energy resources, food, consumer goods and minerals worth over US$ 1,800 million every year. Any challenge to the free flow of oil or goods, which can be interrupted by a host of state and non-state actors, can throw the world economy into turmoil.

The Indian Ocean Region is beset with the full range of security threats – from land and maritime boundary disputes to terrorism and proliferation efforts. Many of the countries either do not have a democratic form of government, or are fragile democracies, which leads to political instability. One of the spin-offs of instability is terrorism, which has been manifesting itself on the high seas too.

This is the broad, sweeping view that one gets of the maritime scene when looking out of the porthole of an Indian warship, and one that defines the context in which the Indian Navy needs to be placed today. Let us now look at some areas, which are cause for concern, either to us or to our friends.

Areas of Concern

Over the past 58 years we have seen nations all round being swept by the floodwaters of totalitarianism, fundamentalism or autocracy. In the midst of this, India has stood firm as a rock to ensure that her people exercise their fundamental democratic right of universal franchise every five years with clockwork regularity. However, it is obviously easier to provide voting rights, than to meet the aspirations for social reform and economic uplift of India's billion strong population in a relatively short period of six decades.

It would be naive to pretend that our democracy is without any flaws, but a vocal press, an independent judiciary and total freedom of expression ensure that there are adequate "safety valves" as well as channels of redressal for any aggrieved citizen of India. It is also our hope that education and economic prosperity in the coming years will reduce population growth, and wipe out poverty and disease.

In some of our states, administrative shortcomings, lack of employment and iniquitous distribution of land/wealth have alienated sections of the population, and tempted youth to take to violence. In the state of Jammu & Kashmir, and in our North-East, problems of this nature have been further aggravated due to intervention by foreign powers in the neighbourhood, and become a full-fledged terrorist movement. This is an unfortunate situation, which is being contained by security forces, and we are confident, will eventually find a resolution at the political level.

The last issue, on which I will dwell on, is related to India's nuclear weapons capability. As I had mentioned earlier, India completely missed the Industrial Revolution of the 18th and 19th centuries because she was a colony and had to be kept as a vast captive market for the finished products of Britain. The two world wars brought few economic or industrial benefits to India, though millions of our soldiers fought on the battlefields of Europe and Africa. An Indian won the Nobel Prize for literature in 1913 and for Physics in 1931, but that was small consolation for lagging so far behind the industrialized world. Therefore, when Independence came, there was a deep-rooted desire to bridge the technological gap and catch up.

Fortunately, in the early days of the Republic, our farsighted leadership put us on the right track for industrialization and scientific development. Our scientists and engineers were confident of their capabilities, and took up the challenge. India's nuclear research programme dates back to 1946 and we exploded our first nuclear device in 1974. There was an idealistic stream of thought in the country, which sought to abjure the nuclear weapon option and to pursue the lofty goal of universal disarmament. For almost two decades, we showed great forbearance, while a national debate raged on the nuclear issue. Our utopian dream however came to an end when we found that we were hemmed-in on two sides by states, which had no qualms about the possession or even the use of nuclear weapons.

In May 1998, India undertook five nuclear tests and declared her plans to put in place a "minimum credible deterrent". By clearly stating her intentions regarding "no first use" and "no use against non-nuclear states" India also made it clear that her arsenal was purely for deterrence, and not for war fighting. This development evoked unfavorable reactions from many quarters, and the USA went to the extent of imposing economic sanctions against us. This was obviously a watershed in Indo-US relations, but like mature democracies, we kept channels of communication open and a dialogue going, and it is heartening to see that today our bilateral relations are not only back on track, but possibly warmer than ever before.

A Prognosis for India

India's Prime Minister Manmohan Singh who is an economist by profession and an open minded liberal of distant vision has articulated an "idea of India". His vision is that of an inclusive, open, multicultural, multiethnic, and multilingual society, which requires India to demonstrate that liberal democracy can deliver development and empower the marginalized.

Complementary to this vision are a set of hardheaded assessments of India's potential in the near term. These clearly indicate that India is on a growth trajectory, which will make her a major player on the world economic and technological scene, and difficult to ignore whether you like it or not. Let me point out a few salient indicators, which have emerged from a recent analysis by the US National Intelligence Council:

- By the year 2032, India's GDP will exceed Japan's, and in the next 15 years the rapidly rising income levels of a middle class of over 300 million will fuel a huge consumption explosion.

- India is investing in basic research in the fields of nano, bio, information and materials technology, and is well placed to become the leader in some of these technologies. Europe is slipping behind and the US will have to increasingly compete to keep its edge.

- With India's gradual integration into the global economy, tens of millions of working age adults will become available for employment in the world labour market.

- As India's economy grows, governments in South-East Asia may move closer to India to help build a potential geopolitical counter-weight to China.

Against this backdrop, it is appropriate to see how Indo-US relations are set to grow in the years ahead.

Indo-US Bilateral Relations

You are all aware that USA and India have a connection going back over 500 years. In 1492, Christopher Columbus set out from Portugal to discover India. Perhaps a minor malfunction in whatever they used for a "Global Positioning System" led him astray, and he landed on America's shores instead. Eventually, he did realize his mistake, but not before he had named the natives of your country as "Indians". That's why they had to be subsequently described as "Red" to distinguish them from the real Indians - that's us!

On a more serious note, as I said earlier, the relations between our two countries have, over the years, seen ups and downs. We have often seen each other through glasses tinted with pre-conceived notions and historical biases. Conventional wisdom seems to tell us that the US and India have too much in common, not to be natural allies. We share a language, a tolerant multi-racial society, and a firm unwavering commitment to democracy. The essence of true democracy lies in an individual's freedom of speech, thought and action. Therefore, it should surprise no one if the world's two largest democracies have different perspectives on world affairs. We do not need to have a complete congruence of views on all issues, but we do need to build strong bridges between our peoples and to capitalise on the manifold values and perceptions that we do share.

Hillary Clinton was in India in February 2005, and this is what she had to say in her keynote address at a conclave. "My favorite acknowledgement of India's place in history is from our great writer Mark Twain, who said, 'India is the cradle of the human race. The birthplace of human speech, the mother of history, the grandmother of legend, and the great-grand mother of tradition.' So India remains all of those. But it is no longer merely a repository of our common culture. It is an active part of our future...regardless of which administration is in power, the US will stand with you as a steadfast friend and equal partner."

After those brave and stirring words, one really cannot ask for more. Fortunately, the fraternity of the seas is universal and sailors speak a common language, regardless of nationality. Not surprisingly therefore, it is our navies that have led the way in bringing the two countries together during the past decade and a half. We have evolved a whole structure for planning and executing an annual bilateral exercise in our waters, named "Malabar", which has gained in professional content and relevance over the years. Today our surface, air and sub-surface units undertake reasonably advanced exercises with a high degree of confidence, and derive a great deal of mutual benefit.

There is a great deal that we can do to enhance the depth and scope of our navy-to-navy relationship. There is a very positive approach in the Pentagon, but one has to accept that the fact that there are two large bureaucracies which get involved in everything that we do. They work on slightly different wavelengths, and we have to allow a certain time lag for that.

In the years ahead, I see a clear scope for us to work together in many areas. In operations at sea we can cooperate in things like disaster relief, anti-piracy and terrorism patrols, and of course, should a UN mandate arise, we would gladly participate.

Friendly contact between the two Services needs to go down to the middle and junior ranks too, and training exchanges are the best way to enhance it. We have some training requirements, especially in carrier aviation where we have sought help. Submarine rescue is another priority area for us, and we are getting close to an agreement on an arrangement with the USN.

The Indian Navy has so far never had the occasion to acquire or operate a major weapon system of US origin. We have now received certain very interesting offers and we are closely examining all aspects, including lifetime support for the systems. Should any of these fructify into a deal, it would indeed open many avenues of cooperation and strengthen relations between the two navies as well as our industries.

India-Japan Maritime Relations

Japan represents for us not just one of the world's largest and most dynamic economies, or one of the most sophisticated Defence Forces, but also a civilization with whom we have had ancient and lasting links.

Historical ties between Indian and Japan go back to the advent of Buddhism into Japan in the 6th century. Ancient Japanese people called India, Tenjiku, which meant a high place above the sky, which is a translation for Sindhu, one of the ancient names of India. The Japanese Shinto myth of creation by the churning of the oceans, as related in the Ko-Ji-Ki and Nihon-ji is also similar to the Hindu fable of creation.

Japan's military and industrial achievements in the early part of the 20th century, the first by an Asian nation, served as an inspiration to Indian freedom fighters and gave them confidence in their struggle against India's British colonial masters. This culminated in the partnership between Netaji Subhas Chandra Bose and Japan with the ultimate aim of obtaining freedom for India.

In recent times, our friendship has been based on the firm conviction that we are both democracies and have shared economic interests. However, despite the fact that we have so much in common, and so few differences, our ancient friendship and affinity has yet to realize its full potential. This is both a matter of regret, as also an opportunity, for there remains immense scope to enhance our bilateral relations.

India's Maritime Interests and Environment

Today, almost 60 years after independence, India stands proud amongst the world community. We are the world's largest democracy where people of different ethnic origins and professing different faiths live in harmony. Our landscape is as diverse as its people

Adapted from a speech delivered at the JMSDF Headquarters, Tokyo, October 4, 2005. The JMSDF Chief Admiral Takashi Saito was present at this meeting.

with the world's highest mountains, the Himalayas in the North, the Thar Desert on our West, tropical rain forests in the East and 7,500 km of coastline along the Arabian Sea, Bay of Bengal and the Indian Ocean in the South.

If we look at the maritime environment specifically, we see that India has an extremely strategic geographical location in a warm water ocean, astride the major shipping lanes, which stretch from the Persian Gulf to the Malacca Strait. A long coastline studded with deep-water ports, a well-endowed EEZ, a rich hinterland and island territories on both seaboards, which straddle vital sea lanes, completes this picture.

The IOR is contiguous to one of the major oil producing areas of the world – the Persian Gulf, from where many industrialized economies like Japan source a majority of their oil supplies. India's own oil consumption is expected to rise rapidly, with the country likely to become the world's single largest importer of oil by 2050. Apart from the energy lifelines that cross the Indian Ocean, major trade routes also lie in this region. Today, 77 per cent of India's trade by value, and over 90 per cent by volume is carried by sea. I am aware that the percentage is almost 99 per cent in the case of Japan. Ensuring the unhindered flow of oil and commerce from this region is, therefore, a major maritime issue for both India and Japan.

The Indian Ocean is virtually a landlocked ocean, with access to it controlled by several choke points, through which shipping has to necessarily pass. These vital choke points need to be kept open at all times to keep both our economy as well as the global economy running smoothly.

Consequently, the area of direct interest to us, which I term as our strategic maritime frontiers, extends from the Persian Gulf, down to the east coast of Africa, and across to the Malacca Strait. Anything that happens in this region has a direct or indirect effect on our maritime interests and security and we need to be not only aware of it but must also be capable of responding appropriately, if required.

We live in uncertain times in a difficult neighbourhood. A scan of the Indian Ocean littoral shows, that with the exception of a few countries, all the others are afflicted with one or more of the ailments of poverty, backwardness, fundamentalism, terrorism or internal insurgency. Many countries are also either ruled by military dictatorships or by authoritarian regimes. Most of the major conflicts since the end of the Cold War have also taken place in or around the IOR. A number of land and maritime boundary disputes in the region linger on, almost all of them being the legacy of the colonial era.

Political instability is directly linked to economic weakness, which in turn, breeds discontent. Not surprisingly then, the fountainhead of terrorism and the trans-national nexus of blatant nuclear and missile technology proliferation are located in our immediate neighbourhood. Dr A.Q. Khan's nuclear Wal-Mart activities are a case in point. These threats pose a major hazard to the world at large. They have maritime dimensions, which will require multilateral efforts to tackle effectively.

Other areas of serious concern are Piracy and Low Intensity Maritime Conflict waged by non-state entities. These cut across state boundaries and require a multilateral response. To counter these crimes, which include piracy, gunrunning, drug smuggling and poaching, the Indian Navy has been engaged in joint patrols along our maritime boundaries with Sri Lanka and Indonesia. In September 2005, another MoU was signed with the Chief of the Royal Thai Navy on the same lines.

All of the above adds up to a fairly intense scenario at sea. The Indian Navy, therefore, needs to possess the necessary wherewithal to counter the entire range of threats that we could confront at sea, and our long-range plans cater for it.

Roles of the Indian Navy

Against the backdrop I have outlined, the Indian Navy sees its involvement in a wide range of operations at sea, extending from nuclear conflict or high intensity war fighting at one end of the spectrum, to humanitarian relief and stable peace at the other. This range of operations has been broken down into four types of roles, promulgated in the Indian Maritime Doctrine that we published last year. These roles, in general terms could be described as Military, Diplomatic, Constabulary and Benign.

The Military role is the primary role of a Navy and is self-explanatory; therefore one need not dwell on it.

The Diplomatic role of the Indian Navy constitutes one of our most important, visible and useful peacetime roles. In the post-Cold War period, particularly after 9/11, our international commitments have increased to an extent that we now need to set up a separate Directorate of Foreign Cooperation. Today, we are working with eighteen navies at different levels.

Our first priority is to assist our smaller and immediate neighbours with training and material aid. At the same time we also desire close relationships with countries further east, particularly Singapore, Malaysia, Indonesia and Vietnam. Needless to say, the diplomatic role offers us tangible results due to the enhancement of mutual confidence and establishment of interoperability to counter transnational threats, and also enables us to work closely in times of disasters or emergencies. In this context, we see Japan as a potential maritime partner of great significance.

This brings me to the next role of the Indian Navy - the benign role. Tasks such as humanitarian aid, disaster relief, SAR, salvage assistance or hydrography are classified under the heading of "benign". The relief operations undertaken in the aftermath of the 2004 tsunami were a classic example of this role.

In this biggest ever humanitarian relief operation undertaken by the Indian Navy, a total of 38 ships, 30 aircraft and about 5,500 personnel were eventually mobilized. The fact that in the midst of our own domestic disaster, we were able to reach out within hours to our neighbours, has received much attention and acclaim worldwide. However, this spontaneous reaction may not have been possible if we did not have an existing level of trust and confidence with these countries, which was evolved through past interaction.

Last, but not the least is the Constabulary role, where forces are employed to enforce law, or some regime established by international mandate. In the Indian context, some of these tasks have been assigned to the Coast Guard. The rescue of the pirated Japanese merchant ship *Alondra Rainbow*, jointly by Indian Navy and Coast Guard ships in 1999 is a live example of this role.

Indo-Japanese Convergence

Since the end of the Cold War, both India and Japan have realized the commonality in our strategic thinking. During the visit to India in May 2005 of, Junichiro Koizumi the Prime Minister of Japan, the Joint Statement issued after the visit stated, inter alia that: "Indian Navy and JMSDF are to enhance cooperation through exchange of views, friendship visits and other similar activities."

We need to enhance bilateral naval cooperation in several areas. Firstly, the field of training, particularly in training exchanges between our young officers, as friendship needs to be built from grassroots level upwards. An example of this is the warm bond I share with Admiral Takashi Saito, the JMSDF Chief due to our interaction at the US Naval War College. In this regard we deeply appreciate the ship riding programme for young officers and the seminar for Next Generation officers being conducted by JMSDF under the aegis of Western Pacific Naval Symposium (WPNS).

Secondly, Japan and India need to develop a partnership in ensuring each other's energy security, as both countries are energy deficient. At present Japan draws a large part of its energy requirement from the Persian Gulf, and in the coming years India could start obtaining energy supplies from Russia's Far East. A number of Japanese ships visit the Indian Ocean in connection with operations in Iraq and Afghanistan and many Indian ships come to the South China Sea. This offers significant opportunity for both sides to interact gainfully in professional areas.

Thirdly, considerable scope also exists for cooperation in the area of defence equipment. The technical sophistication of the Japanese Arms Industry could be meshed with the low labour costs and economies of scale possible in India, to jointly co-produce or maintain defence equipment of common requirement and origin. Finally, vast scope exists for Japan and India to cooperate on issues such as security in the Malacca Strait, counter-proliferation activities, disaster relief, anti-piracy measures, maritime terrorism and sports and adventure activities.

Track II activities, particularly the Tokyo Defence Forum and the Asia-Pacific Naval College Seminar are also excellent forums for exchanging information and perceptions.

Conclusion

By way of conclusion, it can be stated that the emerging strategic scenario in Asia-Pacific presents both opportunities and challenges. Japan and India will be amongst the largest economies in the world in the coming decades and the likelihood of Japan and India assuming larger responsibilities in the UN Security Council is also a distinct possibility. Indo-Japan cooperation is therefore extremely important to maintain the balance of power in the Asia-Pacific region. Both Japan and India have also to live with uncomfortable neighbours - North Korea in the case of Japan and Pakistan in the case of India. All these factors point toward the necessity for a strategic alignment between India and Japan. This alignment can only take place in the maritime arena due to the imperatives of geography.

I am aware that these are exciting times in Japan. You have an invigorated economy under the dynamic leadership of a reformist Prime Minister. India too is poised to take off economically and a strong maritime alliance between both countries will serve as a strong bond of stability and security for the entire region.

For An Indo-Australian Concord

India and Australia have much in common. Sharing the same geographic hemisphere, both our countries are key players in the Asia-Pacific strategic environment, and more specifically, the Indian Ocean Region. We share the common fraternity of the Commonwealth and speak the same language – though in markedly different accents. Both our countries are deeply, committed to a democratic way of life, and have secular, free and open societies, opposed to all fundamentalist, exclusivist and divisive ideologies. And the clincher - we both play cricket!

Another area of commonality is the fact that we were comrades-in-arms, in the first half of the 20th century, in the fight against fascism and imperialism. The Indian Armed Forces contributed over two million troops to World Wars I & II. Indian soldiers fought and died, alongside the ANZACs at Gallipoli. Thousands more died in various battlefields across the globe, from Europe to Africa to South-East Asia. Their sacrifice is a testament, if any were needed of our enduring commitment and willingness to shed blood in defence of freedom and democracy.

Despite a shared past, the attitudes and perceptions of our countries have, inevitably been shaped by their environment and national interests. India seems to present many contradictions and conundrums to most Australians – just as we are often puzzled by your postures and actions. I will need to delve a little into India's past in order to explain how we arrived at where we are today.

India attained independence from two centuries of British rule in 1947, and today she is 59 years old as a modern sovereign republic. However, as a cultural, religious and civilizational entity with an unbroken history, she is perhaps a hundred times as old. Being a colony, we missed the industrial revolution completely, and at Independence, India's founding fathers were faced with many dilemmas and difficult

Adapted from an address delivered at the Australian Defence College, Canberra, May 9, 2006.

choices. One of these was; which model of economy should be chosen to run the fledgling nation?

Our first Prime Minister Jawaharlal Nehru decided that in order to ensure social and economic justice for India's poor masses, we would follow, what was termed a "Socialistic Pattern of Society". This essentially meant a western style liberal democracy, but with state control of the levers of economy and industry. This model also strived to attain early scientific and industrial self-reliance by import substitution.

We therefore became, for the first four decades of our existence, a closed and somewhat stagnant economy. While a great deal of progress did take place in many fields like agriculture, health and education, we later realized that entering the global economic mainstream could have given us a much more accelerated growth. In 1990, an acute financial crisis forced the Government to take some radical steps, and while gingerly opening up the economy to globalization, started the process of privatizing state owned enterprises. Today, India's economy is forecast to become the fourth largest in the world by 2020, and the third largest by 2050.

Along with our erstwhile Socialistic economy, another area, which created a great deal of confusion in the West, was India's policy of "non-alignment". The world, as it existed in the middle of the last century was a sharply polarized one starkly painted black and white by cold warriors like John Foster Dulles. In this environment, "non-alignment" was merely a manifestation of our desire to remain equidistant from the two power blocs that existed. We were certainly not a Communist state, and at the same time, as a newly independent colony we were equally reluctant to join our former imperial masters and become a member of any of the treaty organisations, which bound the Western world. However, many in the West saw our failure to jump on the bandwagon as a sign of bad faith. We ourselves considered it a badge of ideological independence.

As far as Indo-Australian relations are concerned, the divisions of the Cold War did have an inevitable negative impact. Till the early 1970s both our navies used to exercise together with other Commonwealth navies under the Joint Exercise Training (JET) programme. Gradually, this connection faded away and in subsequent years our use of Soviet hardware led to a false perception in the West that we were somehow aligned with the Soviet bloc. In Kim Beazley, the former Australian Defence Minister, we found a consistent critic of India's security perceptions.

With the end of the Cold War there was hope that new vistas of cooperation, would perhaps open, but again, India's nuclear tests in 1998 created another impediment in bilateral relations.

As independence approached, there was a great longing in India to make up for lost time and quickly get on the right track for scientific and industrial development. India's nuclear research programme was launched as far back as 1946; and we exploded our first nuclear device in 1974. However, there was an idealistic stream of thought in the country, which sought to abjure the nuclear weapons option and to pursue the lofty goal of universal disarmament. For almost two decades, we showed great forbearance, while a national debate raged on the nuclear issue. Our Utopian dream, however, came to an end when we found that we were hemmed-in on two sides by states, which had nuclear weapons and openly threatened their use.

In May 1998, India undertook five nuclear tests and declared her plans to put in place a "minimum credible deterrent". By clearly stating her intentions regarding "no first use", "non-use against non-nuclear states" and a "unilateral moratorium on testing" India also made it clear that her nuclear arsenal was purely for deterrence, and not for war fighting.

For many years the West, led by the US, did not quite seem to understand our viewpoint on nuclear weaponisation. The signing of the Indo-US nuclear energy agreement in March 2006 between Prime Minister Manmohan Singh and President George Bush has been a vindication of our principled stand on nuclear issues over the decades. It has now been recognised that India has shown great restraint and moderation in the development of her nuclear arsenal despite having possessed the indigenous capability for over three decades. We have an unblemished record as far as non-proliferation is concerned, which is in stark contrast to many other nations; some in our neighbourhood and others in Europe which have recklessly aided proliferation.

State of bilateral relations

Perhaps the 'Joint Statement' issued after the visit of Prime Minister Howard to India in March 2006 is a good place to begin. Prime Minister Howard put it most succinctly, when he remarked: "There is so much that we have in common, but for a combination of reasons in the past, the potential for cooperation has not been fully liberated. I have the feeling that this potential between our countries is now being freed of earlier constraints..."

From the realization of this potential when both sides identified trade in goods and services, investment, defence, security, education, science and technology, environment, civil aviation and sports as areas of future cooperation. Both sides also discussed ways of rapidly expanding cooperation in counter terrorism. Six agreements, including a MoU on Cooperation in Defence, were signed. Both countries identified the emerging regional architecture as an area of focus for both countries and both committed themselves to strengthen peace and cooperation in the Asia-Pacific region.

We need to look at bilateral relations in terms of both the common threats we face and the opportunities we have for working together.

First, the opportunities. Bilateral trade between our countries is already growing strongly with India being the sixth-largest destination of Australian exports in 2005 worth Australian $ 6 billion, which comprised 5 per cent of Australia's total exports? While this figure in itself may not be very impressive, the significance lies in the fact that it is growing annually at a very healthy rate of 22 per cent.

The trade balance remains squarely in Australia's favour and is mainly centred on the export of raw materials like gold, copper, coal and wool to India. There is tremendous scope for further strengthening this trade by the export of natural gas and uranium. The issue of export of uranium to India is being seriously studied by Australia. There is also a lot of scope for the services sector and this is an area where Indian expertise can be of assistance to Australia.

The other area where I see strong opportunity for growth in our bilateral relationship include collaboration on science and technology issues, where as I mentioned earlier, India is growing strongly while Australia already has substantial expertise. Increasing education opportunities for Indian youth in Australia (about 21,000 students studied in Australia in 2004) and the growing Indian Diaspora down under, which numbers over 150,000 are also serving to bring us together due to increased people-to-people contact. I foresee this as an important driver in our bilateral relationship, much the same way as the Indian Diaspora in the US has played an important role in bringing India and the US closer together. Civil aviation and sports has also been identified as an area of collaboration between both countries.

The second issue, and one that is of relevance to us in the military, is the issue of common threats and how we can work together to combat them. The need to ensure the free and safe flow of seaborne trade is a common objective for both our countries. Both countries are heavily dependant on overseas trade – 83 per cent of Australia's exports and 95 per cent of India's exports by volume are seaborne.

The requirement to counter terrorism in all its forms is another vital security imperative for both countries. India has been, and continues to be, a target of terrorists over the past two decades. Thirty-two innocent villagers were massacred in cold blood in May 2006 by terrorists in Jammu and Kashmir. Three days ago, the Taliban beheaded an Indian engineer working on a road-building project in Afghanistan. I am aware that Australia too has paid a heavy price in the War on Terror.

We have learnt over the past two decades that it does not pay to draw a distinction between terrorists in one part of the world from another. The oft quoted dictum that one man's terrorist is another's freedom fighter is disingenuous and only a ruse by unscrupulous elements to achieve their nefarious ends. The bitter lesson about the horrifying nature of terrorism learnt from the 9/11 and Bali blasts should never be forgotten. To combat this scourge India and Australia had signed a MoU in Combating International Terrorism in 2003, which is designed to forge closer cooperation between our security, intelligence and law enforcement agencies.

With 11 per cent of the world's coastline, Australia faces significant challenges in policing her maritime areas. Since both of us have well established blue water naval forces with a common objective and similar philosophy, it makes sense for both to exercise and operate together in furtherance of their common security interests.

Today, the IN exercise with most major foreign navies, but bilateral interaction with Australia has so far been limited to the exchange of observers to our multilateral MILAN exercise conducted at Port Blair in the Andamans, and your Kakadu exercise. This interaction can be enhanced to ship-level exercises in the future. Another area where we can benefit immensely is in sharing information to enhance our mutual Maritime Domain Awareness, and thereby ease the identification of illegal elements at sea.

Conclusion

It is my conviction that Indo-Australian concord and harmony can only be an influence for good in this region. The realpolitik of the post -9/11 era, combined with the national interests of both nations will drive Indo-Australia relations in the future. The fact that we share the same geo-strategic space and have common threats and interests is of course an added catalyst. People of goodwill in both countries must, therefore, hope that all this, underpinned by our common history and shared values will bring India and Australia closer to each other for mutual benefit. The armed forces of both countries also have a significant role to play in this process.

THE 1,000 SHIP NAVY
A Global Maritime Network

If the daring proposal put forth by the authors strikes a chord in this part of the world, it is for two reasons. Firstly, the Indian economy having emerged from many years of self-imposed isolation has been experiencing the joys as well as the tribulations of globalization for the past decade and a half. And secondly, when terrorism struck a savage blow in the heart of New York on 9/11/2001, we had already been battling this dreaded scourge for many years; in a conflict that extends from our mountains to our seas.

Globalization has led to tremendous growth in seaborne commerce and increasing exploitation of the seas for other purposes. This, in turn, has been inevitably accompanied by an increasing trend towards piracy and other maritime crimes; we thus see the new "silk routes" attracting latter day buccaneers who respect neither national laws nor international boundaries. Man-made borders are also meaningless before natural disasters, and we have seen the devastation wrought by earthquakes, tsunamis, cyclones, floods or droughts cutting across nations and clearly demonstrating that no single country can by itself hope to cope with such catastrophes.

As I write these words, a biennial event named "Milan 2006" is taking place in Port Blair in our Andaman & Nicobar group of islands. Milan is Hindi for "Confluence", and nine navies have gathered here from the "extended" Bay of Bengal neighbourhood to meet, and glean mutual benefits that professional and cultural fraternization bring. This year's

US Naval Institute Proceedings, Volume 132/3/1237, March 2006, p. 41. In the November 2005 issue of this journal, two aides of the US CNO, Admiral Mike Mullen, had mooted an interesting proposal in an article titled: "The 1000 Ship Navy: Building a Global Maritime Network." The above piece was written in response to a request from the Editors of the USNI Proceedings asking for the reactions of Naval Chiefs worldwide.

Milan has special significance since Port Blair and surrounding islands suffered major damage from the tsunami of 2004. Amongst various other activities of Milan 2006, there will be seminars on Maritime Terrorism and Natural Disasters.

Living in a dangerous neighbourhood, and having established bilateral patrols, and initiated regional maritime cooperative endeavours, albeit in a small way, we understand what the authors are talking about. We agree that such cooperation if extended globally will require resources that go well beyond the capability of any single nation or even a group of nations.

The 1,000 Ship Navy mooted by the authors will, however, be an international "force in being" which pre-supposes not just "willing navies" but also political consensus across the board, and reassurances related to sovereignty issues, intelligence sharing, command & control and other transnational concerns. There would also have to be a clear understanding and commitment for contributors to this "dormant navy" to respond when the call comes. In real life however, it is possible that this excellent concept may founder on the rocks and shoals of national sensitivities.

A possible way that I would suggest is to establish the "1,000 ship navy" under the aegis of the United Nations and ask each member to earmark certain "blue-hatted" units. The UN, through its maritime agencies like the IMO, has already demarcated areas of responsibility for hydrographic survey, issue of navigational warnings and search and rescue responsibilities. This format could be replicated or modified for policing purposes, thus maintaining good order at sea, as envisaged by the "1,000 Ship Navy".

Does The Empire Need Help?

With the collapse of the Soviet Union, the United States assumed a position of power unseen for a millennium and a half, since the collapse of Roman Empire. With power comes hubris; and the eternal dilemma of Empire has been how to balance hubris with prudence. Recent US history does not provide very encouraging evidence of their ability to manage this challenge adroitly.

The last great test faced by the US was in Vietnam 45 years ago. Having reduced her World War II adversaries to abject and unconditional surrender, the US was convinced that the collapse of colonial powers had cast the mantle of hegemony on her, and a *Pax Americana* was essential for the benighted nations of Asia and Africa. But the main motivation behind her urge for intervention was the conviction, spurred by Dullesian and McCarthyite rhetoric that Indo-China was the first of a series of "dominoes" that would knock over the others in a chain reaction, if it fell to communism.

Having committed 40 per cent of her combat ready divisions, 50 per cent of her air power and over a third of her navy to Vietnam by 1969, she was forced to beat an ignominious retreat six years later.

As far back as 1961 there were enough straws in the wind, provided by the earlier revolutionary struggles in Algeria, Malaya and Indo-China, for the US State Department to have drawn up an exhaustive list of lessons, as well as policy goals, before plunging headlong into the bottomless pit that was to be Vietnam. But either myopia or hubris; or possibly a combination of both convinced people in positions of responsibility that an unlimited budget, combined with overwhelming military capability would ensure that America's might prevailed. They were to be proved sadly wrong; at the cost of a divided nation and a demoralized military.

Force, Volume 4, Number 9, May 2007. pp. 18-19.

Three decades on, with crack intelligence organizations, and elite think tanks at their disposal, both the US administration and the military failed to gauge the impact of their "shock and awe" regime-change campaign on the Iraqi psyche. The British could have helped here, but they were either not asked or not heard. After all it was they who re-drew the map of the region in the post-Ottoman era, under a League of Nations Mandate. Distinguished British soldiers like Allenby, T. E. Lawrence and John Glubb "Pasha" had fought alongside and against the Arabs, and had intimate knowledge of the Arab mind.

So, are we seeing Vietnam redux in Iraq? With the benefit of hindsight one give the answer in affirmative. With possibly worse to follow.

As a nation not known for historic recall or strategic foresight, India can hardly afford to gloat over US discomfiture. On the contrary, we have cause for serious concern because any worsening of the situation in Iraq or Afghanistan, or a conflagration in neighbouring Iran could have a disastrous impact on our promising, but fragile economy, and shatter our dreams of prosperity forever. As it is, we have paid dearly in terms of national security while complacently watching the sinister proliferation nexus between Pakistan, China and North Korea make free use of the sea routes for exchange of missile and nuclear hardware for years.

It is for this reason that India's attitude of detachment, with regard to developments in the neighbourhood, is worrisome. Unless we are involved, we will have no leverage, and unless we have some leverage, we are powerless to influence the course of events vital to our national security. Nepal, Sri Lanka and Bangladesh are three glaring examples in our volatile neighbourhood where we have chosen to remain hands-off, but where things could blow up in our face overnight, and catch us unprepared, because we have no "trip-wires" in place.

India's armed forces too, have had a long association (mainly in the training arena) with both Iraq and Afghanistan. However, we have a fundamental difference with the US here. On principle, India does not participate in any initiative, which is not conducted under the UN ensign. And the US, which is loath to place any forces under UN Command, has been going great lengths to avoid it.

But there are signs that things may be changing. In mid-February 2007, I was invited to be a "Senior Mentor" for a course conducted in Hawaii, by the Commander US Pacific Fleet for 23 Navy and Marine Corps Flag Officers. The objective of the course was to prepare these senior officers (which included an Indian Commodore, a Sri Lankan Rear Admiral, and four other officers from South-East Asian navies) to take command of a Coalition maritime force in operations "other than war".

The participants received comprehensive briefings/presentations regarding a range of legal, operational and administrative aspects of multi-national operations at sea, and undertook an exercise in planning and organizing a coalition task force for a hypothetical disaster relief scenario. I was struck by three aspects.

First of all, there was a surprising volte face. The settings for the classroom exercise opened with a UN General Assembly Resolution, which taking note of a tsunami in the Indian Ocean, and welcoming the US offer to lead a multi-national force, "authorized the deployment of such a force for relief to be coordinated by the UN Office for Humanitarian Assistance."

Secondly, in a talk by the US Navy Chief Admiral Mullen, mention was made of the concept of a "1000 Ship Navy" which recognizes that a shrinking US Navy no longer has the wherewithal to act or intervene worldwide, and seeks the cooperation of like-minded nations in forming this notional maritime force to collaborate in countering maritime terrorism, piracy and rendering humanitarian assistance.

And finally, all dissenting observations, whether they related to unilateral intervention, non-compliant boarding of ships on the high seas, or placing the "1000 Ship Navy" under UN aegis were received in a very positive spirit by all US officers present, who appeared to welcome and applaud multilateralism. In private, senior USN officers also rued the increasing US tendency to act unilaterally, which often left them out on a limb.

The US sees itself today, as a world leader which can set an international agenda, but the snowballing violence in Iraq, and the see-sawing military situation in Afghanistan have perhaps brought home the fact that it is spread thin, and not strong enough, by itself, to implement such an agenda world wide. In the light of this experience, it is possible that the US may be now coming around to the gradual acceptance of multi-polarity, and the restoration of multilateral military cooperation, under the aegis of the United Nations.

It is not that we have not worked with the US forces before. Soon after 9/11, in a show of solidarity with the US in the global war against terrorism, India committed two helicopter equipped offshore patrol vessels carrying Marine Commandos to Operation "Sagittarius". For a five-month period between April and September 2002, the IN provided escort for 24 "high value" units of the US Navy, including nuclear submarines, during their transit through the Malacca Straits, under Operation "Sagittarius".

The frosty relationships of the Cold War era are now a distant memory, and the "level of comfort" between the armed forces of the two nations has multiplied manifold over the past decade and a half. Frequent bilateral exercises have created not just a fund of goodwill, but also deep and genuine professional respect for each other. During the torturous negotiations preceding the passage of the Henry-Hyde Nuclear Cooperation Act, a tacit understanding had emerged between the military leadership on both sides; it was agreed that regardless of the outcome of the parleys, they would stay the course, and not allow the gains – painstakingly made – to be frittered away on account of diplomatic pique.

The navies of the two nations – for reasons that need no elaboration – have been at the forefront of many initiatives. In April 2007, India's CNS hosted his counterpart, the US CNO, and they had an agenda packed with issues of mutual interest. It is perhaps a fortuitous coincident that the Eastern Fleet of the Indian Navy sailed for the Pacific on an operational deployment unprecedented in scope, duration and professional content. One of the highlights of this deployment was the unique tripartite exercise between the IN, USN and the Japanese Maritime Self-Defence Force. Our ships then proceeded to Vladivostok for a bilateral with the Pacific Fleet of the Russian Navy.

This may be a good time for India's security planners to step back and take a hard look at the big picture through a "Palmerstonian" prism. Keeping our long-term vital national interests foremost, can we find convergence of views with the US, and thereby place ourselves in a position to influence events in our part of the world? If the Empire does indeed need help, should India contemplate stepping in to lend a hand – under the aegis of the UN, of course?

The US would be immensely relieved to receive valuable support from a respected regional power. The Indian Armed Forces would be happy to obtain priceless operational exposure in areas of strategic interest to us. And India's national security would gain immeasurably if we acquire leverage in our "near abroad", a extended neighbourhood.

SPEECHES AT COMMISSIONINGS AND CEREMONIAL OCCASIONS

THE PRESIDENT'S FLEET REVIEW 2006

All of us assembled here are witness to a unique and momentous occasion for the Indian Navy. The President of India, the Supreme Commander of the Armed Forces, Dr A.P.J. Abdul Kalam, is in our midst and will break bread with us. In the next two days, the President will review the combined fleets of the Indian Navy anchored off Visakhapatnam and then he will present the Colour to the Eastern Fleet on February 13, 2006.

This is the first time since Independence, that the Navy has decided to hold the Fleet Review off the East coast, and it has significance for more reasons than one. First, the Review coincides with a shift in focus of India's attention to her eastern neighbours, which makes this venue indeed more relevant and appropriate. The President's highly successful recent tour of the Far East was a manifestation of the nation's "look east" policy. This is reflected in the maritime re-awakening of our eastern seaboard; and the port city of Visakhapatnam, is at the heart of this process. The Navy owes much to Vizag, and it is our sincere hope that the Fleet Review will contribute to its prosperity and well being.

The second, and more important factor is historical. In ancient India, while the west coast saw mainly commercial seafaring activity, our eastern waters were the medium for intense maritime interaction with South-East Asia. Chandragupta Maurya's empire, which ruled the shores of the Bay of Bengal in the 3rd century BC, was the first imperial power in our history. His minister Kautilya has chronicled in the Arthashastra, detailed instructions and guidance for managing a seagoing fleet. The Mauryas were followed by the Satavahanas, Pallavas, Chalukyas and Cholas, all great seafaring dynasties, which ruled peninsular India and sent out fleets that carried India's trade, culture and religions to South-East Asia. And finally came the great Sri Vijaya dynasty, which founded an empire in Sumatra.

Adapted from a speech delivered at a ceremonial banquet held in honour of the President of India, Visakhapatnam, February 11, 2006.

India's cultural bequest to this region is still visible in Angkor Vat, in Borobudur, in Ayuthia, and on the islands of Java and Bali, among other places. And it all came from our eastern shores. The fall of the Sri Vijayas brought about the decline of India's maritime power in the 13th century, and this coincided with our domination by foreign powers for the next seven centuries. But even during this interregnum, the east coast continued to witness significant events of maritime history.

During the late 18th century when the struggle for control of India was on, the French Admiral Andre de Suffren, in alliance with Hyder Ali, waged a brilliant campaign against the East India Company's Fleet. Suffren fought and defeated the British Admiral Hughes in a series of sea battles off Madras, Pondicherry, Cuddalore, Nagapattanam and Trincomalee. Had France shown greater maritime vision and supported Suffren, India's history might have taken a different turn.

We were touched by the First World War, when the German cruiser *Emden* was sent all the way from the Atlantic in 1915 to bombard Madras, much to the consternation of the British. In World War II, after the Japanese occupation of the Andaman Islands in 1942, an invasion of the east coast was considered imminent, and the British constructed many Naval Air Stations on the southern peninsula, one of them in Vizag. In 1943, the British aircraft carrier HMS *Hermes* was sunk just outside Trincomalee harbour and her wreck can still be seen on a clear day, in about 30 metres of water.

And of course, the 1971 Indo-Pakistan war saw a glorious chapter of India's maritime history being written in these waters. The newly formed Eastern Fleet of the Indian Navy under the command of Rear Admiral S.H. Sarma, was deployed for operations in the waters off East Pakistan. Led by the carrier INS *Vikrant*, the Eastern Fleet mounted a bold and determined blockade off the enemy coast. Our carrier-borne aircraft interdicted Pakistani airfields, ports and shipping, and our warships cut off both reinforcements as well as escape routes of Pakistani forces. The domination of the Bay of Bengal by the Eastern Fleet exerted immense pressure from seawards and expedited the surrender of East Pakistan to Indian forces, on December 16, 1971.

In the 35 years that have elapsed since, the Eastern Fleet has not only grown in size and combat power, but has also shouldered much greater responsibilities in the maritime security of our eastern waters. The presentation of the President's Colour to the Eastern Fleet is thus a well deserved recognition, and an acknowledgement of the contribution made by this frontline combat formation of the Indian Navy.

There is a vital linkage between our country's economic well-being, national security and maritime power. India's resurgence is dependant on seaborne trade and energy security. It demands that we have a navy, which is commensurate with the country's vast maritime interests and regional status. Today we have a ship-building programme, which will deliver, in the next decade and a half, ten fast attack craft, four anti-submarine corvettes, six offshore patrol vessels, three landing ships, six destroyers, six submarines, and an aircraft carrier. I cannot think of any navy in the world today, which can boast of comparable acquisition plans.

Our Supreme Commander has done the Navy proud by consenting to spend with us, 48 precious hours during which we hope to provide him a macro as will as micro view of how the men and women of the Navy operate and work, especially when he makes a historic dive on a submarine the day after tomorrow.

In the US when they want to give a huge compliment to someone, they say he is a "rocket scientist." Dr Kalam is certainly a "rocket scientist", but he is also much more. He is a scholar, poet, musician, a teacher and a person who is sensitive to the problems of the common Indian. When the President reviews the fleets, we are sure he will feel a sense of pride at the sight of Indian–built warships all round him. Much of the equipment on board comes from the laboratories that he once headed. This is a preview of tomorrow's navy, which we hope will be totally designed and built in India.

ON MILITARY LEADERSHIP–I

I want to highlight a few aspects of the naval service and military leadership, which will stand all the Naval Cadets passing out in good stead during their service in the Indian Navy.

The first thing to remember is that, you do not merely work; you serve. Belonging to the Armed Forces is not the same as any other job because to serve implies that you should be ready to sacrifice your time, effort, comfort and even your life for the sake of your country. The Navy is not a 9-5 job, but a 24-hour, 365-day commitment.

The next issue is of responsibility, accountability and authority. Once you go on board your ships on completion of your training, you will have to shoulder enormous responsibilities. But remember, that with authority and responsibility comes accountability. In a few years, some of you may find yourselves as the officer of the watch on the bridge of an aircraft carrier, or in the conning tower of a submarine, or in the cockpit of an aircraft with crores worth of equipment and a large number of men under your charge. To discharge your duties will require a high degree of professional knowledge and commitment, as you will be accountable for all your actions and inactions.

The third issue is trust. The Navy operates on trust. The watch-keeping system of the Navy is based on trust. Those on watch are entrusted with the safety and well-being of their ship and their shipmates. Remember the Navy will forgive an honest professional mistake; it will never forgive a breach of trust.

A few words about military leadership. The first and most important attribute of leadership is to know your job. Remember, in the close confines of a ship, submarine or aircraft, your knowledge or your ignorance will be known to one and all in a short period. If you are found wanting, you men will not trust you and your leadership will be ineffective.

Adapted from the speech delivered at the passing out parade of the 69th Naval Academy Course at INS Mandovi, Goa, November 27, 2004.

In today's high-tech world, learning is a constant process and success will go to those who constantly update their knowledge.

The second essential attribute of a good military leader is physical and moral courage. While physical courage is easy to understand, moral courage is a more complicated issue. Simply stated, moral courage implies the ability to stand by your beliefs and to choose the harder right from the easier wrong. Both physical and moral courage is born out of strength of character and a sense of belonging and patriotism – both to the service as well as the nation. A shining example of this courage was the stoic willingness with which Captain M.N. Mulla, MVC, went down with INS *Khukri* on December 9, 1971. By offering his life jacket to one of his subordinates, and calmly supervising the evacuation of the ship he exemplified all the virtues that I have mentioned – physical courage, for he knew that death was certain; moral courage, for as Captain he took full responsibility for the event; and concern for the safety of his men.

The third attribute of good leadership is teamwork. On board a ship everyone has to work and fight as a team. Unless you can foster team spirit amongst those whom you command, your performance, both in peace and war, you will never rise above the average. To achieve team spirit, apart from the attributes of physical and moral courage and professional excellence, is required a profound regard for, and knowledge of your men. I can quote no finer example than that of Lord Nelson, whose intimate knowledge of his men made mere mortals into heroes.

As you leave the portals of this Academy, you also need to be very clear about the vital national role of the service that you are joining. Over 90 per cent of our trade and a large proportion of our energy needs are met by imports carried by ships. We have tremendous natural wealth and resources in the seas as well as under the seabed. The economic well-being and future prosperity of our great nation is, therefore, inextricably linked with the Indian Navy's ability to safeguard our interests at sea.

The Indian Navy is also a force for peace and stability in the oceanic regions around us. Our ships are often seen in the ports of our neighbours to render assistance, provide re-assurance or just to extend hand of friendship. Every officer and sailor of the Indian Navy conducts himself as an ambassador of the country to spread the message of goodwill and harmony. From today, those of you who pass out will be a part this representation and will have to conduct yourselves accordingly.

On Military Leadership–II

It is undoubtedly a special day for all the Gentlemen Cadets, who stand at the threshold of a new phase in their lives. It is also a day of joy and happiness for the parents and guardians, whether they are present here or not, and I congratulate them for having sent their wards to join the Indian Army. For the staff of the Academy, there must be a feeling of deep satisfaction, for they would have laboured hard to impart quality training, to mould the young leaders of the future.

You have reason to be proud, because you are joining the ranks of one of the finest fighting forces in the world: the Indian Army. In a short while from now, you will assume positions of responsibility in your units across the length and breadth of our country, and start making your contribution to the security of the nation.

From now onwards, the support, the resources, the facilities, and the brotherhood of not just the Army, but of the Armed Forces are yours to rely upon. To this select fraternity, I extend you a warm welcome today. But if you think that your newly acquired status guarantees you a life of ease and comfort, you are mistaken. In the course of your service, you will often face hardship, loneliness, discomfort, and danger.

Sooner or later in your careers, you will wonder what the compensations for this hard life are. Let me tell you. First of all, you will have the rare privilege of holding the President's Commission, proudly wearing the uniform of your regiment, and serving your country. And should you be so fortunate, you may even go into battle to defend the honour and integrity of your Motherland. Secondly, being an officer will cast you in the role of a leader – and this is the most important and unique privilege that any young person can ask for.

Adapted from speech delivered at the Combined passing out parade at the Indian Military Academy, Dehradun, June 11, 2005.

From now onwards, you will have the honour and privilege of commanding the finest fighting men in the world - the Indian Jawans. Wherever you serve, you will have the loyalty, camaraderie and devotion of these magnificent men. These are the men through whom you will accomplish your mission. It is, therefore, your sacred duty to ensure that you are worthy, in every respect, of the respect and devotion of the men that you will command. And in this regard the words of Field Marshal Chetwode, engraved on the portals of the IMA, should be your guiding light.

Remember, that while your men are duty bound to obey your orders, this does not mean that they will automatically accept you as a leader. While the President's commission can be conferred upon you, the mark of leadership has to be earned through sweat and toil, and by proving yourself to your men. Since leadership is a key quality for an officer, I want just mention a few factors, which have a bearing on this vital requirement.

First among them is personal example. If you have the ability to show your men exactly how you want something done, by doing it yourself, there will seldom be need for persuasion or compulsion. Your example will itself act as the greatest inspiration and motivation for your men. Leadership by personal example may be difficult, but it remains the most effective way of leading men - both officers and jawans.

The second important requirement of leadership, particularly at your young stage in service, is that of physical courage. The Indian Army has a long history of valour, but in the recent past there can no better example than the raw physical courage displayed by young officers on the forbidding heights of Kargil. On daunting terrain, in bitter cold, and often under withering enemy fire, they led right from the front. Many of them laid down their lives for the country's honour, and set us an example worthy of emulation.

The third vital ingredient of leadership is moral courage. Moral courage means the strength of character to distinguish right from wrong, and the wisdom to always choose the harder right over the easier wrong. Remember a wrong is always a wrong, no matter how small it may seem. Both physical and moral courage are born out of strength of character and a sense of belonging and patriotism.

The fourth important requirement is of professional knowledge. You are in the business of soldiering, and your men will gauge your standing very quickly. If you inspire confidence in them about your professional competence, they will follow you unquestioningly.

Finally, one of the qualities we speak a lot about in the Armed Forces is loyalty. It is something that you must reflect upon later, but there are only two types of loyalty, which I would commend to you. One of course, is loyalty to the subordinates that you will command, and the other even more important quality is loyalty to your Service. From your first day as an officer, you must pledge that no matter what the price, and it could well be your own career or advancement, you will remain loyal to the Army, always place the Service before your personal interests, and never do anything that will sully its fair name.

If you can do all this, you will have achieved the moral superiority, necessary to lead your men into battle, and even order them to stake their lives.

As you leave the portals of the Indian Military Academy, you must do so confident in the knowledge that you have received all the inputs necessary to take your rightful place in this fine Service that you have decided to make your career. From this day onwards, the Army will continuously enrich you in every way; with training, and guidance, with experience, and exposure and by placing invaluable opportunities before you. I am confident that you will repay your debt to the Service and to the nation, with your loyalty, dedication, hard work and patriotism.

On Military Leadership—III

It gives me great pleasure to be here today, to review the Combined Graduation Parade of cadets who leave the portals of this fine institution on completion of their training. I offer all of you, my warmest felicitations.

As you stand on the threshold of a new phase of your lives, this day assumes a very special significance. It is also a day of joy and happiness for all parents and guardians. Whether they are here or not, I congratulate them for having sent their wards to join the Indian Air Force. For the Academy staff, there is reason for deep satisfaction and pride. They have labored hard to impart quality training, in the air and on the ground, to mould the young leaders of our air force, and they too deserve felicitations.

As a naval aviator, I have had a long and happy association with the Indian Air Force (IAF), going back to the days of my youth. About 37 years ago, I received my wings not very far from here at what was then the Fighter Training Wing, Hakimpet. The Indian Air Force (IAF) has contributed in substantial measure to what I am and where I stand today; and this is an appropriate occasion to acknowledge my personal gratitude to this fine Service.

The nation looks towards the IAF not just to defend its skies and airspace against all threats, but also to carry the war into the enemy's territory should the need arise. Air power is a formidable instrument of offence and defence, and you are all going to play a vital role in its employment.

If you happen to be flying an aircraft, you may be in the front line, and have your moments of glory. If you are in a support service, remember you are helping to keep the aircraft flying, and their crews to achieve their mission. An equal share of the glory therefore belongs to you. No matter where you serve, and in what capacity, or what

---

Adapted from the speech delivered at the Passing Out Parade at the IAF Academy, Dundigal, June 18, 2005.

badge you wear on your chest; as an IAF officer, each one of you will be making a vital and unique contribution to the nation's security.

Your contribution will commence, in a short while from now, when you assume positions of responsibility in the IAF. You will be blessed with the gift of comradeship of your fellow officers; and you will have the precious loyalty of the men you command. Challenge and adventure will become a part of your daily life. From now onwards, the support, the resources, the facilities, and the brotherhood of the not just the Air Force, but of the entire Armed Forces are there for you to rely on.

To this select fraternity, I extend to you, a warm welcome today. But if you think that your newly acquired status guarantees you a life of ease and comfort, you are mistaken. In the course of your careers, there will certainly be occasions when you face hardship, loneliness, discomfort, and danger. We recall with great respect and admiration, the memory of those who lost their lives in the Battle of Kargil, and in IAF Station, Car Nicobar during the tsunami. You will be expected to take such things in your stride, because as an air force officer you will place the interests of your country, your Service, your unit, and the men you command, well ahead of your own.

Sometime or the other, in the course of your career, you will wonder what are the compensations? Let me tell you. First of all, you will have the great privilege of holding the President's Commission, the pride of wearing the smart blue uniform of the IAF, and serving your country. Should you be so fortunate, and the call to arms comes, you will have the honour of participating in the defence of your Motherland. And finally, wearing the rank badge of an officer will cast you - no matter what your branch - in the role of a leader. And this is the most important and unique privilege that any young man or woman can ask for.

No matter where you serve, in India or abroad, or which unit you serve in, it will be your loyal, professionally competent and trustworthy airmen who will ensure that your mission is accomplished successfully. To command them is a privilege that no other profession can offer.

The mantle of leadership does not come easily, and has to be earned by gaining the respect of those you command. Since this is an attribute, which will play an important part in your careers from now on, let me mention a few factors, which will help you become better leaders.

First among them is personal example. Never ask your officers or men to do anything that you will not happily and gladly do yourself, all the time. On the other hand, if you lead the way by your own personal example, there will seldom be need for persuasion or compulsion. My most abiding memory of the 1971 war is the sight of my squadron commander leading the first mission over enemy territory. His example served as the greatest motivator for the entire squadron: officers and airmen, throughout the war.

The second important requirement is of professional knowledge. Your men will gauge your professional standing very quickly. The IAF today deploys some of the most sophisticated aircraft, weapons and sensors in the world. Operating, maintaining and supporting them, requires professional knowledge of a very high order. But I am referring, not just to hardware. Whether it is the art of administration, financial management skills, or the mysteries of meteorology, you will need to be on top of your profession to be a professional officer and a credible leader.

The third vital ingredient of leadership is moral courage. This implies the strength of character to distinguish right from wrong, and the wisdom to always choose the harder right over the easier wrong. In an unfair world, sometimes there may be a price to pay for showing moral courage, but that is part of the bargain.

The fourth attribute of good leadership is the ability to inspire teamwork. While a pilot is up in the air, there is an entire supporting cast of maintainers, ATC officers, Fighter Controllers, Met officers and administrative staff who underwrite his effort. Without such support, he may not have got airborne in the first place. You must develop the ability to foster cohesiveness and unity amongst your team; so that your endeavours are blessed with success in peace and in war.

Finally, one of the qualities we speak a lot about in the Armed Forces is loyalty. There are only two types of loyalty, which I would ask you to imbibe and nurture. One is loyalty to the subordinates that you will command, and to the comrades who serve alongside you. The other, more important quality is loyalty to your Service. From the first day in the Service, you must pledge, that no matter what the price and it could well be your own career or advancement, you will remain loyal to the Air Force, and never, ever, by word or action sully its fair name.

As you leave the portals of the Air Force Academy, you must do so confident in the knowledge that you have received all the inputs necessary to take your rightful place in this fine Service. From this day onwards, the Air Force will continuously enrich you in every way; with training, and guidance, with experience, and exposure and by placing invaluable opportunities before you. I am confident that you will repay your debt to the Service and to the nation, with your loyalty, dedication, hard work and patriotism.

May you always touch the skies with glory!

ATTRIBUTES OF A STAFF OFFICER

To most of us, before we come to the Defence Services Staff College, (DSSC) Wellington it appears like some distant Shangri La, a kind of wonderland where the climate is always salubrious and life is one big party. Students are quickly disabused of this notion in the first few months of their existence in Wellington, but gradually the schedule eases out, you make friends and start enjoying the experience with your families. Needless to say, the parting memories of this wonderful institution are invariably pleasant and so the legend of DSSC lives on.

Well, your time at Wellington is up. I am sure that a profound sense of nostalgia has already set in, even though you would be keenly looking forward to your new assignments. You will shortly carry the Staff College stamp - the *"psc"* after your name. It is an onerous responsibility, for your superiors and subordinates will expect you to discharge your duties and behave in a manner befitting a Staff College graduate. The Directing Staff would have spent the last 45 weeks instilling the qualities required to discharge your future duties, both as a staff officer and as a commander.

The fact that you were chosen for the staff course is indicative of your potential for higher command. Now that you have completed your course, what is it that will be expected of you? Very broadly, your future job content can be divided into broadly two categories – combat functions and peacetime duties, and I will touch upon both of these categories. First, your combat functions.

You are aware that the origin of military staffs, as we know them today, lies in the Prussian Army of the late 18th and early 19th centuries. It is illustrative to note that the Prussian Army under Frederick the Great in 1740 comprised just 85,000 troops and even then was the fourth largest standing army in the world. The growing complexity of

---

Adapted from the Valedictory Address delivered during the Graduation Ceremony of the 61st Staff Course at the Defence Services Staff College, Wellington, April 28, 2006.

modem armies required a dedicated body of professionals to manage issues such as planning, intelligence and logistics. The Prussian model of selecting the best officers to carry out these 'staff' functions led to the adoption of the staff officer model, first by the French Grand Armee under Napoleon, and subsequently by the other military powers of that time. The creation of staff colleges and a specialized cadre of officers to advise the commander on various functions in war followed. Over two centuries after the adoption of the concept of the 'staff officer' his job definition in war has not changed too much.

War is primarily concerned with two sorts of activity – the delivering of energy and the communication of information. In the Vietnam War, the United States delivered to Indo-China enough energy to move 3.4 billion cubic yards of earth – ten times the amount dug out for the Panama and Suez canals combined. It does not require the intellect of a rocket scientist to realize that if this energy had been properly directed, the outcome might have been different. The second issue is that of communication of information, and is equally complex. In combat, even with the best modem technology, danger, uncertainty and the fog of war can combine to create communication road-blocks, which can result in lethal consequences. The famous, but misconceived Charge of the Light Brigade is an example that I can offer in this context.

Therefore, in a combat situation, a good staff officer is one who can first, correctly plan and advise his commander on where he most gainfully needs to deliver his offensive energy; and second, communicate his commander's intent most clearly and efficiently to the subordinate commanders.

In peacetime, while the dangers and uncertainties of war will not hang above your head like the proverbial 'Sword of Damocles', there will be other pressures to contend with. You will be expected to give the 'staff opinion' on a variety of issues, from selection of a missile to deciding on the methodology of rehabilitation of war widows. You can bet your bottom dollar that this will always be required yesterday if not the day before.

To be a good staff officer and a leader, you need several attributes, which I will now mention.

First and foremost is the "Knowledge of the Technique of Making War" or professional knowledge as we call it. Unless you are a thorough professional, you will be of little benefit as a staff officer. But professional knowledge alone is insufficient. You will be well advised to read as widely as you can - particularly history and geo-politics.

The next invaluable attribute of a staff officer is the "Ability to See the Few Essentials to Success", or clarity of thought. In an era of instantaneous communication and excessive information, time will always be in short supply. A good staff officer is one who can discriminate between the essential and the superfluous. Your opinion will always be more valued if you can offer well-reasoned solutions in crisp language. This is where correct judgment and decision-making, which the Staff College would have introduced you to, will come into play. More than your superiors, this quality will be valued even more by your subordinates, particularly in crisis situations. Remember, an indecisive leader is even more of a liability than one who is ignorant.

You will need to have the "Courage of your Convictions", or the ability to withstand pressure. This implies giving your honest professional opinion, without fear or favour, and putting it down in black and white. It does not mean toeing your superior's line or situating the appreciation, in Staff College lingo. I can assure you that if you are honest, correct and measured in your opinion, your counsel will be valued. On the other hand, a sycophant is of little use to any organisation, and will soon be marginalized.

A "Well-balanced Judgment", is another very essential attribute. Both as a staff officer and leader, you need to have the ability to understand your subordinates, separate the chaff from the wheat, and choose the best man for the job at hand. Sometimes, in a crisis, this will require an instantaneous decision, which can only come if you have a close knowledge of human nature. So, it is very important that you have a good understanding of human psychology.

I will just mention one last attribute, and that is the "Ability to Inspire Team Work". You can never win with a one-man army. It requires a team to win. To be able to inspire team work you will need to trust your subordinates and above all, have faith in your men and your cause.

A word on jointness. The DSSC is one of the premier 'Joint' institutions of the Indian Armed Forces. I sincerely hope that over the past year you have not just imbibed jointness in a cosmetic form, but have understood its essence. It is a truism that no major operation over the past century has ever been executed by an individual service. In the years ahead, the need for synergy between the three Services will be inescapable. I can also foresee that if we do not move towards more jointness ourselves, it is likely to be thrust upon us by the environment. I hope you have used your time at the Staff College well to understand each other's strengths and limitations, as you will need them in the years ahead.

COMMISSIONING OF INS *KADAMBA*

It is a naval tradition to formally commission all our units, whether afloat or ashore, as "Indian Naval Ships," with due nautical ceremonial. The commissioning of INS *Kadamba*, by the Defence Minister Pranab Mukherjee, and the christening of the base by the UPA Chairperson, Sonia Gandhi marks the completion of the first phase of our new base at Karwar, and this depot ship from now onwards support and coordinate all the activities that take place here.

Before moving further, I would like to mention that it was Admiral Dawson, a former Navy chief, a visionary who conceived this project and selected this location. His presence amongst us is an honour for all of us.

This is an occasion of tremendous significance. The inauguration of a sprawling, modem, state-of-the art base, meant exclusively for the Navy's use is a remarkable event; not just in the annals of our young navy, but also in the maritime history of modem India. From this base will sally forth, warships and submarines flying the Indian white ensign to stretch out the hand of friendship to our neighbours, to safeguard our vast maritime interests, and should the need arise, to deliver a lethal blow to the nation's enemies: at sea, or from the sea.

It is not often that a nation invests precious resources to create a bastion of national security on a scale such as this. It is also a matter of rare good fortune that so many men of vision happened to be associated with a project of such importance. The most prominent supporter of this concept of a third naval base was our former Prime Minister Rajiv Gandhi, who took pains to come all the way in October 1986, to lay its foundation stone.

I have no doubt that the location of this base will prove to be of great benefit to the people of this region, and it is my hope that the Indian Navy will make a valuable contribution to general development of Karwar.

Adapted from the speech delivered at INS Kadama, Karwar, May 31, 2005.

The Navy and the nation have waited for a long time to see this base coming to fruition. Although this project received approval in principle in 1985, a resource crunch led not only to delays, but also to a scaling down of the original plan. It was only in 1995, that a truncated Phase I was approved, whose construction commenced in the year 2000. This ceremony marks the culmination of this first phase of construction.

At the heart of this base will be the Naval Ship Repair Yard and a unique ship-lift facility, which would provide technical support to ships based here. These are also nearing completion, and we hope that by the end of this year we would be able to commission them. For this however, we have to overcome a manpower ban and to persuade the Ministry of Finance to accord sanction for the civilian manpower to run it.

A lot of time and effort has also gone into ensuring that a comprehensive compensatory package was evolved and implemented for the rehabilitation of all the local inhabitants who had to move because of the project. Apart from the re-settlement package, one person from each affected family has been guaranteed a job in the naval base. I am sure that the presence of this base with its sophisticated technical facilities will not only provide a boost to the local economy, but also open many avenues of employment for local youth.

Project Seabird remains till today, the most ambitious infrastructural undertaking of the IN. Such a complex and massive project could not have been completed successfully without proper leadership. This is perhaps the appropriate occasion to pay a tribute, and to acknowledge with gratitude and admiration, the sterling contribution of everyone who worked towards the realization of this dream.

The responsibility of operationalizing this base now rests upon the first Commanding Officer of INS *Kadamba*, and his commissioning crew. We all know that pioneers always have a challenging task, but blazing a new trail is an immensely satisfying experience. I am confident that the commissioning crew will not only execute their task with zeal and dedication, but also set the highest standards while laying a sound foundation for this base. Do remember that those who go to sea are our first concern, and the Fleet always comes first.

COMMISSIONING OF INS *BEAS*

It is with a sense of immense satisfaction that I stand here today to commission the third, and the last ship of Project-16A: INS *Beas*. It is a time-honoured naval custom that old ships are never allowed to die. After they are retired or de-commissioned, new ships inherit their name and the tradition thus carries on. There are present here, ten people, including myself, who had the privilege in an era bygone, of commanding the first ship named *Beas*. For all of us, it is a matter of great good fortune to witness the old *Beas* come to life again in a new and more powerful incarnation.

At independence, we inherited a fleet of assorted old warships, to which we added a few more second hand acquisitions. The *Beas* was one of a batch of eight brand new warships that were built in British yards for us and delivered in the early 1960s. In those days of course, we had no choice but to buy everything that we needed, from abroad. It was the vision and foresight of our predecessors that, in the face of much skepticism, set us on the path of indigenous warship building.

This potent Indian-built warship that stands before you represents not just reach and striking power, which is many times more than that of her first namesake, but also the proud achievement of our shipbuilding industry. Today, we can legitimately claim to be a 'builder's Navy' and I learn with great pride, which every Indian will share, that INS *Beas* has the highest-ever indigenous content of any warship built in India thus far.

It is indeed a unique and emotional occasion for me to commission the reincarnation of a ship that I was privileged to command a quarter of a century ago. It is not often that such an event takes place in the active service of a naval officer, and I consider it a singular honour. Another happy coincidence is the fact that the Commanding Officer of the ship being commissioned today, Captain Jamwal, happened to be a young midshipman

Speech delivered at the commissioning of INS Beas at GRSE, Kolkata, July 11, 2005.

on board the old *Beas,* under my command in 1981. The baton is thus passed on from one generation to the next in the extended family that is the Indian Navy.

The commissioning of INS *Beas* marks the successful completion of the Navy's strategic plan to start a second frigate construction line. Today we have two shipyards in the country, capable of undertaking the construction of frigate sized warships of considerable sophistication: the Mazagon Docks on the West coast and GRSE on the East.

For GRSE, the attainment of new skills involved in building ships of this type has not been easy, but their resolve and determination has paid off. *Beas* is being commissioned just a year after her sister ship *Betwa;* which in itself is a considerable achievement. This fine ship that you see here is an embodiment of the professional skills, fine workmanship, commitment, and dedication of each and every member of this yard's team, and they all deserve to be warmly complimented.

The commissioning of *Beas* also marks the completion of a series of 12 steam-propelled ships of a common origin; and this programme also highlights the history of our ship-building post-Independence. Commencing in the late 1960s, with the construction of the *Nilgiri* class frigates, which were essentially license built ships of the British *Leander* design; our Naval architects took their first independent steps. They designed the *Taragiri* class of ships, which were a development of the *Leander* and carried a large helicopter. The next step in this progression was the broad-beam *Godavari* class, which were missile armed, and amongst the few ships capable of operating two large helicopters. The *Brahmaputra* class, to which this ship belongs, is the culmination of this unique design evolution and a landmark for our designers as well as shipbuilders.

Over the years, we have designed and built several classes of ships like the Delhi class destroyers, the *Magar* class landing ships, and the *Khukri* and *Kora* class corvettes. INS *Mumbai* took part in the Royal Fleet Review in Portsmouth UK, and I have already received many compliments that she was the smartest and most impressive ship in that international gathering of warships. The fact that the Indian Navy's Design Organisation and our shipyards are presently involved in the design and production of five new classes of ships, including the Indigenous Aircraft Carrier, is an indication of how far we have proceeded down the path of a national capability in warship-building.

The ship standing proudly before us is fine testimony to our current prowess in indigenous warship design and construction. Her impressive array of weapons and sensors, and many other features embodying cutting edge technology enable her to compare with the best in the world. She also represents a capability which unique to Indian shipbuilding: that of blending and interfacing technology of Russian, Western and Indian origin to work efficiently together. It is for this reason that our ships evoke wonder and admiration wherever they go in the world.

The Indian Navy today bears the responsibility of safeguarding the nation's vast and vital maritime interests, and maintaining peace and tranquility in the seas around us for the common good. India's growing trade and economy, and the inherent genius of our people are carving out, for this nation a new place, high in the world order.

In the years ahead, India's inevitable growth in stature, combined with our democratic, liberal and secular credentials will ensure that we become the focus of regional and international responsibilities. In such circumstances, there is no doubt in my mind that the Indian Navy will have an increasing role to play, not just the maritime arena, but as a powerful instrument of state policy. Our prompt and resolute action in the face of last year's tsunami provided a fine demonstration of the reach and capability of maritime forces as much in peace, as in war. The requirement of a balanced three-dimensional blue-water maritime force is, clearly recognised by our government and the induction of ships like INS *Beas* is yet another step in the attainment of this goal.

I am glad to inform this audience that the Navy's ship construction and acquisition programme has gathered the requisite momentum. Work on the Indigenous Aircraft Carrier is well underway in the Cochin Shipyard while in the Mazagon Docks, ships of Projects 17 and 15-A are already at an advanced stage of construction. Work on Offshore Patrol Vessels at Goa Shipyard is also scheduled to commence in the near future. In Kolkata, construction of three Landing Ships (Tank), Anti-submarine Warfare corvettes and Fast Attack Craft will ensure that the order books of GRSE remain full for the foreseeable future.

The Indian Navy has firmly held to its belief in "self-reliance through indigenisation." To this end, we have given our full support to the indigenous shipbuilding industry and also allocated funds from our own budget for the modernization of our shipyards, including GRSE. We recognise with clarity, that besides ensuring self-reliance, indigenous shipbuilding also helps to provide a tremendous boost to a host of ancillary industries, enhances the technological capability of the nation, assures employment to thousands and adds to economic growth.

While the ongoing construction activity is certainly satisfying, there is little reason for us to rest on our laurels. To merely sustain the Navy's current force level, we need to induct five to six new ships or submarines annually. It has been our experience that defence PSUs and other shipyards in the country can at best; deliver only 2 to 3 ships per year. This is just not good enough, and our shipyards must do whatever is necessary to accelerate the pace of production. While we are deeply committed to our PSUs, we just cannot allow our force levels to decline further. Therefore, if left with no alternative, we shall have no choice but to import warships to make up our force levels.

The current international standard for building a frigate-sized ship is about 36 months and our shipyards need to ensure that their productivity and efficiency picks up to match shipyards in China, South Korea, and Japan. In fact, not only must GRSE and our other yards be capable of meeting the shipbuilding demands of the Navy, but must they also need to become competitive to undertake export orders. Only then can they really achieve cost-effectiveness.

I must state emphatically that the time has come for us to start thinking seriously of mobilizing the private sector in support of the national ship-building effort. In the Navy we recognise and acknowledge the need to keep our PSUs fully engaged and occupied. At the same time, we find that our private sector has tremendous capabilities, which remain largely untapped and unexplored. While the public sector has its own strengths, there is also much to be learnt from the private sector in terms of efficiency, and productivity, as well as financial and human resource management techniques. The answer lies in synergizing our national resources through the medium of joint ventures, and private-public partnerships. That is the way the rest of the world is going, and we urgently need to follow suit.

I am given to understand that *Beas* has achieved a high level of operational readiness right at commissioning. This speaks very highly of the concerted efforts by many naval agencies and the shipyard.

I am confident that *Beas,* along with her sister ships *Brahmaputra* and *Betwa,* will make a most significant contribution to India's blue water capabilities in the years to come. Today INS *Beas* begins her commissioned service under India's proud white ensign. Let me remind the Commanding Officer and crew of this fine ship, of the proud lineage that they inherit, and express my confidence that they will carry forward the fine traditions of the old *Beas.* It is my sincere hope that this ship will serve the navy and the country faithfully and gallantly, in peace as well as in war. Should you have the good fortune to sail into battle, it is the Navy's expectation that INS *Beas* will vanquish the country's foes and make us proud of her.

COMMISSIONING OF THE FIRST UAV SQUADRON

A little less than 53 years ago today, the first aviation unit of the Indian Navy was commissioned right here in this air station, then known as INS *Venduruthy II*, in March 1953. This was the Fleet Requirement Unit (or FRU), equipped with the twin-engined Sealand amphibians, which used to operate from the Cochin channel.

Since that historic day, the Navy's aviation arm has made tremendous strides, and honed its skills in carrier operations, long-range maritime patrol, ship-borne anti-submarine warfare, and airborne early warning. Currently every major ship of the fleet carries an integral helicopter, which is not only her eyes and ears, but actually a force multiplier for the Fleet.

Our navy is growing in stature and capability to fulfil its role of safeguarding the nation's maritime interests. Naval aviation has kept pace in every way, and it now constitutes a potent arm of the Indian Navy with a long reach, and the ability to combat threats in all three dimensions.

Today we have added a new capability to naval aviation, and to the Indian Navy.

The seas around the Indian peninsula are the scene of intense shipping activity. A high proportion of the world's trade in oil, liquefied gas and other vital commodities transit through the waters around us. Even in peacetime, any threat to shipping or disruption of commerce can cause a major financial crisis worldwide. It is, therefore, essential that peace and tranquility should prevail on the sea lanes. Threats at sea can take many forms: piracy, gun and drug-running, maritime terrorism or just accidental pollution. To maintain order at sea, it is imperative that we have a clear and continuous picture of what is happening in the waters around us at any time. Maritime reconnaissance is, therefore, of paramount importance to the navy; as much in peace as during hostilities.

Adapted from speech while commissioning INS Garuda, Kochi, May 2005.

Today we are at a landmark juncture for both the Navy and its air arm. With the commissioning of Indian Naval Air Squadron 342, we are amongst the first few navies in the world to have inducted unmanned platforms into the maritime scenario. Unmanned aerial vehicles represent a number of advanced technologies; small efficient aero-engines, light composite airframes, miniaturized sensor payloads, and high speed digital data-links. As a Service we are proud to have assimilated them with ease. Flying a UAV also requires special skills, and our officers have mastered them too.

UAVs bring many advantages with them. They are compact, and relatively silent and, therefore, stealthy. They can be sent in harm's way because no human lives are at risk. They have long endurance and are economical. Above all; a UAVs sensors provide continuous information in real time, which often surpasses the quality and accuracy of that provided by bigger, and more expensive manned aircraft.

We were one of the first navies to operate UAVs out at sea in a tropical environment, involving heavy rainfall. It took time to gain experience and expertise, but after three years of intensive flying trials, we are now amongst the pioneers in the esoteric art of UAV operations at sea.

Equipped with the Searcher Mark II and the Heron UAVs, INAS 342 is going to be an asset which will enhance our maritime domain awareness manifold. Capable of operating from any shore location, and being controlled from specially equipped ships, these UAVs will undertake reconnaissance hundreds of miles out at sea. With their optical, radar or electronic warfare payloads nothing will remain hidden from them.

THE 28TH SURANJAN DAS DINNER

I am delighted to be back in here after a lapse of nearly three decades. I recall that when I passed out of ASTE with the 5th PTP Course in 1976, I was a worried man. Because I had about ten years service, and out of that, less than half was with the Navy, and I used to wonder if NHQ would one day transfer me to the IAF for good. Anyway, I was posted from here to the naval equivalent of a Base Repair Depot in Cochin and did some production test flying. After six months we received the news that the Super Constellation squadron was to be taken over by the Navy, and sure enough my name was on the list of the commissioning crew.

Till then I had considered myself a hard core "fighter jock", and NHQ seemed to be offering me the ultimate insult! So for the first time in my life, I put in an application for redressal of grievances complaining about the offence to my *amour propre!* I was informed that since I had flown the Canberra, HF-24, and the HS-748 during the PTP Course, I was one of the few pilots in the navy with twin-engine experience, so I had no choice but to convert to the Super Connie. For quite some time, I hated ASTE for doing this to me. Till I realized that the Super Connie wasn't a bad machine, and it was nice to have a Flight Signaler bring you hot coffee now and then!

Before I go any further, I would like to say a few words in remembrance of the great aviation legend in whose honour this dinner is named.

Born in 1920, Group Captain Suranjan Das set off to be an engineer, but at the outbreak of World War II, decided that what he actually wanted to do was fly. He qualified as a pilot in Canada, and was commissioned into the RIAF in 1943. Having established a reputation for solving tough flying problems, he gravitated towards test flying, and was sent to ETPS in 1949. On return, he was deputed to HAL where he joined the HT-2

Adapted from the Valedictory address delivered at the passing out of the 28th Test Pilot's Course at the Aircraft & Systems Testing Establishment (ASTE), Bangalore, May, 26, 2006.

development programme. He was thereafter associated with the assembly and acceptance flights of the Ouragan's shipped out from France, before proceeding to Europe with a team to evaluate the Mystere, Hunter, Canberra and Gnat for the IAF.

With the adoption of the Gnat by the IAF, Gp Capt Das was attached to the Folland works at Chilbolton in UK and did a great deal of developmental flying on the aircraft including the first ever demonstration at Farnborough by an Indian. However, his greatest achievement by far in Chilbolton was the acquisition of a lovely wife! On return to India, he continued to contribute to the safe induction of the Gnat into IAF service, and in the process, modifications were incorporated in the gun firing and tail plane systems of this mini fighter.

Gp Capt Das then entered the HF-24 project, and spent considerable time test flying various versions of this unique aircraft. He was engaged in testing a re-heat version when on January 10, 1970; he met with an unfortunate accident and lost his life. He was posthumously awarded the Padma Vibhushan for his services to Indian aviation. He combined in himself all the qualities that one would look for in an officer, a gentleman and a courageous test pilot, and it is only befitting that this dinner commemorates his memory.

After almost a year of hard labour, you are now poised on the threshold of an exciting and challenging vocation. From here on there will not be many dull days in your calendar. You will now have a job in which you will constantly strive to stay abreast of technology. Your task will be made more difficult by the furious pace of progress in various fields of aerospace science, which is matched only by the rapid rate of obsolescence. A clear understanding of the design philosophy behind the systems you are testing will always be essential; to tackle unforeseen eventualities that will often face you in test flying.

However, while dealing with the arcane, you must never lose touch with the pedantic and mundane. Remember that the "Joe Pilot" in the field will have neither your razor sharp and analytic mind, nor your lightning reflexes. And yet he is the customer, and it is his interests that you must jealously safeguard. Do not ever forget the days when you cursed the aircraft designer for an awkward feature of the aircraft, or author of the Pilots' Notes for some impracticable check. Now it may be you who will write the Pilots Notes or can change a design feature.

As a Test Pilot you will form the vital link between the designer and manufacturer at one end, and the user at the other. The rapid growth of the Indian aviation industry has spawned several prototype aircraft, which are flying in our skies today. Computers and flight control systems notwithstanding, it will be your job to bridge the gap between the designer's ambitions and the user's requirements as well as his safety. To us in the Navy, this dilemma is quite familiar, as we try to interface two machines: the ALH and the LCA designed for land-based flying with the demands of ship-borne operations.

ASTE is a unique institution since it not only caters to the test and evaluation needs of our Armed Forces, but also generates expertise, vitally required by our aviation industry and DRDO. The success of our future aviation projects depends upon the expertise of personnel who will evaluate them, and there is a need to ensure that graduates of the Air Force Test Pilots School (AFTPS) acquire the necessary expertise and exposure. Only then will it be possible to ensure timely and successful completion of the ambitious projects we have embarked upon.

I am happy to learn that Air Force Test Pilots School is keeping pace with the changing technology and consistently upgrading its training facilities. I understand that in a commendable initiative, the Test Pilots School, in collaboration with Visvesvaraya Technological University, has started an M Tech degree programme in "Flight Test Engineering". The flight test engineers graduating today are the first batch of engineers who will be awarded Masters Degree after one year of additional work. A similar programme is being worked out with Bangalore University to award M Sc degree in "Flight Dynamics" to the test pilots in the subsequent years. I am also happy to learn that the School is shifting into a new modern building and the next course will be conducted from a custom built abode.

Another step forward is its affiliation to the Society of Experimental Test Pilots and Society of Flight Test Engineers, the IAF Test Pilots School has gained immensely in stature. This is yet one more milestone in your journey to attain international status and recognition. In this arduous journey, the contribution made, by our own research and industrial institutions over the years also needs to be acknowledged. From my own days in ASTE, I recall the ungrudging support that has been traditionally rendered by HAL, ADA, NAL, ADE, LRDE, and GTRE; and I express gratitude to them on your behalf.

The Indian Navy has always been alive to the need of qualified test crew and has maintained a close interaction with ASTE and AFTPS since their early days when I was a student here, along with the Deputy Chief of Air Staff. We have always cherished our relationship with this elite aviation institution and nineteen Naval test crew comprising three PTPs, five Fixed Wing test pilots, nine Rotary Wing test pilots and two flight test engineers have so far graduated from AFTPS. I am happy to note that the twentieth test crew is graduating today.

Out at sea, naval aviation as you know operates in a somewhat unique domain, especially during ship-borne operations. The maritime flying environment presents challenges both from the piloting as well as the maintenance angles, and many of these require the application of skills and experience that only a test pilot can bring to bear. As the repository of tremendous aviation expertise we look to the ASTE for support in many of our new projects.

The most important of these is the LCA (Navy). In trying to adapt a tail-less, delta wing fighter for a ski-jump take off and an arrested landing, ADA has bravely taken on a major challenge. We plan to operate the LCA from our aircraft carriers in the next decade, and are in the process of setting up extensive flight-testing facilities in Goa. In this project, we look forward to the full cooperation, advice and assistance from ASTE.

On our part, I reiterate the commitment of Indian Navy to fully support the flight-testing activities at ASTE and the training at AFTPS. The Navy has actively been involved in providing support for evaluation of SU-30 MK I and the radar of the maritime Jaguar as well as for test exercises on Naval Kamov/Sea King helicopters and the Harrier simulator. We understand and appreciate these collaborative endeavours as they foster "Jointmanship" and enrich the common fund of aviation expertise.

I want to say this to the young test pilots and flight test engineers who have graduated this evening. Flying is by itself a demanding, exciting and satisfying vocation. By choosing flight-testing as your specialty, you have entered the ranks of an elite which sets the standards for others; both in aviation as well as in industry.

This will require you to carefully cultivate certain qualities in yourselves. You will of course need flying skills of the highest order to meet every eventuality in the air. You will also need courage; not just the physical variety to handle emergencies, but also of the moral kind to back up your convictions to the hilt. And above all, you must retain at all costs, your professional integrity, no matter what kind of pressures that you may have to face. In all this you will have a role model before you: the legendary Test Pilot; Gp Capt Suranjan Das, in whose memory this graduation ceremony is named.

HUMAN RELATIONS

A Matter of Honour

Many of us have probably never heard of Jeremy Boorda. He was an American citizen of a nation whose ethics we continuously sneer at, from the high moral pedestal of India's ancient civilization and culture.

I start this article with a mention of Boorda, because on May 23, 1996 he drove home from his office, drew a loaded pistol and ended his life by shooting himself in the chest.

Admiral Boorda was the four-star head of the US Navy, known as Chief of Naval Operations (CNO). The 56-year-old Flag Officer was the first man in the history of USN to rise from the ranks to the post of CNO. In an hierarchy dominated by US Naval Academy graduates, Boorda's humble origins meant that he had to prove himself at every Ring of the ladder in his long climb to the coveted post of CNO. So what made him take his own life when he had reached the pinnacle of professional achievement? Was it cowardice, or was it a sense of honour?

For a number of years, Admiral Boorda had worn a combat insignia (the small metallic letter "V") affixed to two campaign ribbons earned for shipboard service off Vietnam between 1965 and 1973. As per USN regulations, this insignia was to be worn only by personnel who were deployed in specified combat zones, and a routine check by the Bureau of Personnel had revealed in 1987 that Boorda was not entitled to wear it. By the time that he became CNO, Boorda had stopped wearing the Combat "V", but in April 1996, this issue had been raked up by *Newsweek* magazine, which had sought an interview with him to discuss it. The interview was scheduled the day Boorda shot himself.

Were two tiny pieces of metal stuck on scraps of ribbon, as a result of what Boorda himself termed, "a honest mistake" enough for him to take his life? Opinions will certainly differ, but Boorda had always stressed the long US Navy tradition of leaders accepting responsibility and accountability. He himself wanted to be seen as such a leader, and

USI Journal, January-March 1999. pp. 3-14.

apparently could not tolerate the idea that his personal actions might dishonour the service he had joined at the age of 16.

"Death before dishonour" has been the creed of warriors for centuries. Roman legionnaires, Rajput princes, Japanese samurais and British officers have lived and died of this credo. Some British regiments invoked it even off the battlefield. An officer considered guilty of a serious misdemeanour by his peers would one evening enter his quarters to find a loaded revolver in the opened drawer of his writing table. The unstated message from his comrades was stark but unmistakable: "Do not sully the name of Regiment by a messy Court martial. Take the honourable way out". Very often he did.

To return to Boorda's case, there can be little doubt, that with some legal advice and hair splitting, he could have proved that he had done no wrong and clung to office. It is, however, obvious that he must have agonized hard over this, and come to the conclusion that his service was more important than him, or his life.

The NDA's Dilemma

In the current Indian environment a credo like "Death before dishonour" would appear melodramatic, and perhaps even comic. But, at least for the Armed Forces, there is another credo, more down to earth, and certainly within everyone's reach: "Service Before Self", roughly translated into Sanskrit as "Seva Parmo Dharma". It is the motto of the National Defence Academy that I have had the honour to command, and which has produced a significant proportion of the officers in the Armed Forces today.

An analysis of the backgrounds of the 600 or so teenagers who enter the National Defence Academy annually shows that they come from every walk of life and every social strata imaginable. They are highly motivated young men who cope with the rigours of training better than we did over three decades ago, considering that the curriculum has become far more demanding and intensive now. This is an excellent portent for the egalitarian and highly professional Armed Forces of our Republic. However, it becomes clear from a scrutiny of the range and scale of misdemeanours, which occur in the Academy, that a large proportion of these young men have received no inputs about a value system, nor were they provided a moral foundation at home, or in school.

As a direct consequence of this absence of any ethical moorings, many of them, in the high-pressure training environment, tend to fall prey to the urgings of their more worldly-wise and less scrupulous seniors. They are told that a "smart" cadet should possess the basic "skills" to prevail, or at least survive in any adverse situation. Some of the measures recommended in this unwritten "survival manual" include lying, cheating, manhandling, stealing (or 'management' as it is euphemistically termed), and impersonation. While it may be understandable for a rudderless young cadet to eagerly grasp such concepts, what astonished me was the benign and even approving attitudes of some Divisional Officers and Squadron Commanders towards such grave infractions or what were once sacred traditions of the Academy. A little reflection showed that coming from the same environment, and being products of exactly the same system, these relatively junior officers did not know any better, and hence say nothing wrong in approving something they had experienced themselves during training.

An indoctrination campaign, and some preventive as well as stringent punitive measures have helped reduce, to a great extent, the incidence of such offences. However, it was felt that the current social environment demanded that the young man in the Academy be provided with a tangible code of conduct which would spell out clearly what he was expected or not expected to do.

The Academy Honour Code

The Honour Code systems followed in military academies were studied and dissected for weaknesses. A thick musty old file containing notes and enclosures spread over a quarter of a century (which had repeatedly considered and discarded this concept) was examined and discussed. What finally emerged was the NDA Honour Code Promulgated in March 1998 and a system of administering it.

The Academy Honour Code will, in a short while, be one-year-old. Despite skepticism all round (except surprisingly amongst the Cadets) it shows signs of taking firm root. This code (if it works) will see a cadet through his basic training, and perhaps even later in life. But what is there to guide our officers, especially in the upper reaches of the military hierarchy, when they falter and stumble?

Self Before the Service?

The year 1998 was a trauma filled one for the Indian Armed Force as a whole, and the IN and IAF in particular. The events of this period will be analysed for a long time to come. But if one goes deep enough, and takes a holistic view of series of dismal episodes that one witnessed with a sense of surrealism, the underlying case becomes quite apparent: failure of leadership. Or to be more specific, failure of senior officers to place the larger interests of their Service before personal or parochial interests. Once considered a rare or unusual phenomenon and even dishonourable, taking recourse to courts of law in order to obtain redressal for service related grievances has become commonplace today. Only the shortsighted will fail to discern that the increasing intervention of the courts in what should be internal affairs of the Armed Forces will destroy their spirit, cohesion and morale, in a matter of few years. There are two very good reasons for this.

First, whenever an officer wants to represent his own case in a convincing manner in a court of law, it is inevitable that he will have to show either the Service or his brother officer(s), or both in a poor light. And, this amounts, to nothing else but washing the family's dirty linen in full public gaze. The officer will also feel the need to enlist support of bureaucrats, politicians and the press to bolster his case; in the process, demeaning himself and his Service further.

Second, the judiciary, are a set of hardheaded professionals whose job is to weigh the evidence presented before them, in as dispassionate and impartial a manner as possible. Therefore, they can be expected to disregard any pleas based on abstract notions like customs and traditions of service, command responsibility, esprit de corps, or morale of the Armed Forces. Hence, it is quite likely that most judicial decisions given purely on points of law will violate these notions which are an integral component and lifeblood of any fighting force.

A Hard Look at Ourselves

Yet it is the stand of many officers, that they were driven to litigation only as last resort, in order to obtain redress against "Injustice". Or if we want our officers to foreswear courts of law, it is obvious that we have to aim for two basic objectives. First, every officer must be taught from a very young age (perhaps as a Cadet) to have a clear understanding of what constitutes the interest of his Service, as distinct from what is self-interest. He must also be taught to always hold the former well above the latter. This may perhaps modify his perception of "injustice" in later years. Secondly, we must ensure that our personnel management systems are equitable and fair, to ensure that injustice is not actually inflicted on any one.

It would be naive and simplistic to imagine that one can either pinpoint a single fault-line or suggest a specific nostrum to bring about instant peace and harmony in the services. However, if we can somehow make a few small but significant changes in the way many of us at senior levels do things some of these ills would disappear.

I would like to touch upon a few sensitive issues including some of our common failings here. I do so without attempting to strike a sanctimonious note, because some of the follies and shortcomings related here may be my own!

Overweening Ambition

Ambition is a highly desirable trait in a human being, and especially a fighting man. Without it, there would be no aspiration for higher things, no quest for perfection, and a person could well become an uninspired cabbage.

Having said that, I must add that everyone, and especially fighting men must guard against "overweening ambition". And ambition becomes "overweening" when you start putting your personal advancement above all other consideration. Nothing remains sacred before such an all consuming passion; friends can be stabbed in the back, the service can be shown in a poor light, and all the means including the press, politicians and courts enlisted for furthering ones personal agenda.

In order to retain our sense of proportion where ambition is concerned, we have to keep reminding ourselves of just two basic facts. Firstly, that our service is bigger and more precious than all of us, and must take priority over our personal needs in every instance. Secondly, let none of us delude ourselves that he must embark on a holy crusade because he was "destined" to become a GOC, a Fleet Commander, and AOC-in-C or whatever post he covets. If "A" does not get this post, "B" or "C" will; and whoever gets it will probably do as good if not a better job than "A" would have.

The Resignation Option

Having tried all means of redressal within the service, if one still feels aggrieved, is there any alternative to litigation and the ignominy that it brings to all parties?

Yes, indeed there is! In fact, in the good old days, when going to courts was still considered a sordid thing to do, the honourable option was to "put in your papers" or resign. It is a sign of the times that this option is not considered very "smart" these days. I know of one or two officers who did resign from service on issues of principle, and they are still remembered with great respect and affection by juniors and contemporaries alike. You cannot say the same about those who take recourse to writ petitions.

Frankly, the pensionary benefits that a senior officer gets today are most generous, and the peace of mind as well as the self respect which come with a graceful retirement are probably a million times more precious than the extra promotion or some years of service one may wrench out of the service through the law courts.

Creating Coteries

Sycophancy is a two-way transaction and the burden of guilt must be shared as much by the junior who butters up a senior, as by the senior who encourages or even permits such blandishments. Sycophancy possibly comes more easily to Indians because our culture in any case demands respect for age and position, and it is very easy to blur the very fine line dividing the two.

If not ruthlessly crushed by those of us in senior positions, sycophancy inevitably leads to the formation of cliques and coteries. And this is how it all starts: a junior officer will come up to you one day and say (he may perhaps write), "I want to express my gratitude to your Sir, for the promotion/medal/ posting/course, you have got for me. I know how difficult it must have been and I am really thankful to you".

At this juncture, all you have to do is to shrug your shoulders modestly and makes an innocuous remark like "Oh, its all right, it was no big deal". You may or may not have been responsible for his bonanza, but by accepting the credit, you would have made him your slave for life. This man will forever sing your praises, and you will now feel obliged to "take care" of him. The "quid pro quo" would have been established.

In such a case, my recommended line of action is to seat the person, and then administer a severe rebuke to this effect: "Old boy, if you need to actually thank someone for your promotion/medal/posting/course. You obviously did not deserve it, and we should have it cancelled. On the other hand, if you have really slogged to earn it, why give away the credit you deserve yourself by giving thanks to someone?"

By this line of action, you may well deprive yourself of a potential admirer and slave, but you would certainly have nipped a sycophant in the bud and hopefully prevented the start of a coterie.

One may well ask, what is so bad about a senior officer having a small coterie of bright young officers who share his views and ideas on most things, and admire him? The short answer is, that coteries create an unending cycle of sycophancy and patronage, which manifests itself as described in the paragraphs that follow.

Show me the Face.....

The cynical young officer in a typical ship's Wardroom (and I presume it is so in regimental messes and squadron crew rooms too) often sums up our personnel policies with the cynical phrase: "Show me the face, and I will show you the rule".

For a personnel management system to be accused of arbitrariness, "ad hocism", and inconsistency is the severest condemnation that it can be subjected to. While some of these charges may arise from faulty perceptions, a majority of them are founded on fact. Because goal posts are actually often shifted to dispense patronage, and norms changed to include or exclude people, depending on whether they are in this camp or that. The worst sin that can be committed in this domain is attempting to form a line of succession, and then re-framing the rules to ensure it.

Fairness, transparency and consistency are the best way of inspiring confidence, making coteries redundant, and minimizing instance of perceived injustice. Then perhaps, people will not feel the need to go to court.

Listen to Advice

Contrary to the popular notion, nether age nor rank invest a senior officer with any special Solomon-like wisdom. They give him only experience, which helps him to tide over many a crisis, which might stump a younger man.

In order to ensure that any gaps in his experience are plugged, and the best advice and assistance is always available to a commander, a complete hierarchy is placed at his disposal. However, in order to give himself the maximum benefit of their expertise, the Commander needs to have an open mind, to welcome new ideas, and even to accept occasionally that he may be wrong.

It is the staff, which should first be receiving what are hopefully the most authentic inputs. It is their job to know or to find out, what the junior officers think, what kind of "baat cheet" goes on in *langars*, mess decks and airmen's messes. It is then, incumbent upon them to brief the boss honestly and accurately - the bad news first and good news later.

The catch here is, that many of us keep our mind as well as doors closed, and thereby shut off valuable inputs. We also make it known deliberately and unconsciously that bad news and contrary views are unwelcome. This breeds a set of courtiers who always bring good news and never contradict the boss. A senior officer who surrounds himself with such people isolates himself dangerously and will certainly take wrong decisions, which may cause resentment, and harm the service. Since his "feed back loop" is impaired, he will never I come to know what he has done wrong, and may continue to compound his follies till they reach serious proportions.

Even in an undemocratic set up like the Armed Forces, seeking a consensus, and taking people along (in policy making) is not a bad thing. It may prevent the senior officer from making a serious error of judgment, and will ensure that the policy once promulgated is sincerely implemented, even by those who succeed him in office.

Conclusion

What do those in the rank and file of the Armed Forces think of what they read and see in the media about the contretemps at the highest levels? What are they supposed to make of the allegations and counter allegations being hurled around? And how do they react when our ladies jump into the fray?

The fact of the matter is, that all this is not supposed to happen, and will not happen if the senior officers of the Armed Forces come to a tacit and unanimous understanding on three main issues which could constitute an unwritten and self-imposed code of conduct.

- Firstly, that our personnel-related policies will be above-board, fair and consistent, and that we will do our best to ensure that no injustice is done to any one.

- Secondly that we will discourage sycophancy and not collect coteries around us. Nor will we show undue favour or bias towards anyone, and we will allow merit alone to count.

- Thirdly, we will, as a matter of honour, forswear the use of courts, press and other external means, to seek redress for grievances, and we will confine ourselves to service channels only for this purpose.

The time has also come now, for the Services to form in-house tribunals (on the lines of CAT) to examine and take decision on grievances of Armed Forces personnel.

Finally the question arises, that if this code of agenda is voluntary of self-imposed (it cannot be otherwise) how do we enforce it?

The obvious answer is, by peer pressure or ostracism. In our rural society, fellow felling is embodied in the act of all families drawing water from a common well, and by the men-folk passing the same hookah from hand to hand when they gather in the evenings. The most serious punishment that can be meted out to a delinquent in the village is exclusion from both these activities, or "Hookah pani band", as it is called.

How about some "Hookah pani band" in the Armed Forces?

What Constitutes Success in Life?

I am well aware that NDA cadets do not regard with great favour, after dinner speeches. It was the same in my times too and that is one thing, which has probably not changed.

I know that after long, hard day, a hearty dinner, and the exciting prospect of "Zero DLTGH" *(Days Left to Go Home, a popular NDA acronym)* tomorrow are the right catalysts for a sleepy audience in the Academy. But right now, I do need a few minutes of your wide awake attention to impart some advice to you on the subject of; "What Constitutes Success in Life".

I have chosen this topic for a good reason. The more I reflect on my years in service, the more I am convinced that the root of most problems in life lies in our inadequate understanding, or a misinterpretation, of what really constitutes success. This understanding is very important because there is a direct linkage between an individual's perception of 'success', and his consequent level of happiness and satisfaction in life. For future officers of the Armed Forces, these are important ingredients of morale and motivation, which in turn, are the key indicators of an effective fighting force.

As schoolboys, and later as cadets in NDA, your success was easy to understand and measure. You were ranked as per your performance in various tests, which were common to all. There were tangible benchmarks against which you could compare your performance. However, as you now embark upon your journey into the wide world outside, life will become infinitely more complicated. And success, when achieved, may often seem both amorphous and very difficult to measure, especially if you do not clearly define it for yourself.

People find that success does not always bring with it the feeling of accomplishment or happiness that they thought would inevitably accompany achievement. To make things

---

Adapted from speech delivered at the NDA, Khadakvasla, on the occasion of the traditional Dinner night held in honour of the passing out course, May 30, 2006.

worse, there is always a tendency to compare your success with that of your peers, and often your own success seems to pale in comparison. Inevitably then, feelings of dissatisfaction start to nag you. So at the threshold of your professional career it is worth spending a few minutes to think what success does or should, mean to you.

They say that to become a successful man in any profession three things are absolutely necessary, a persevering nature, a capacity for hard work and dedication towards your chosen career. As far as the Armed Forces are concerned there are many other attributes necessary, but I would choose integrity, professionalism, and the right attitude as three other essential ingredients, because the lack of any one of these constitutes a major handicap. Remember that Lady Luck too plays a major role in all our lives.

There are no magic formulae for success, so never, for a moment imagine, that if you follow some set of golden rules and always do the right things, success will fall into your lap. There will inevitably be setbacks and moments of despair. It has been said: "He that has never known adversity is only half acquainted with others, and with himself. Constant success shows us only one side of the world because it surrounds us with friends, who will tell us only our merits, and silences those enemies, from whom alone we can learn our defects." So remember, that failure and adversity too can form the stepping stones to success.

Getting the formulae and *mantras* for success is easy. The difficult part lies in determining what really constitutes success. Is success in life to be measured in terms of the promotions that you achieve? Is your success a measure of the financial prosperity that you can attain? Does it lie in the popular acclaim by your family and friends? Or is it something else that is more profound and intangible?

In the Bhagvad Gita, Lord Krishna tells Arjuna: "He who does his duty without caring for its fruit is a true Yogi." Not every one of us can aspire to be a "Karma Yogi", but if you think carefully, you will realize that while we may not have the power to influence the outcome of events, we do have complete control over our own actions: what we can or cannot do. So a good definition of success that you could adapt for yourself is simply: "doing your duty to the best of your ability". Never mind the results that you may or may not achieve.

By this definition, you will have been a successful man if you yourself are satisfied that you have been sincere and loyal to your country, your organisation, your family, and ultimately, to yourself. It would be utopian for me to suggest that you should not aspire for advancement and prosperity. Do so by all means, but do it through the right means and through just efforts.

During your career, your quest for excellence and your ambition will spur you on to great endeavours. That is how it is meant to be; never be satisfied with the "good" if you can attain the "best". But unless you have a clear understanding of what constitutes "success", your ambition, could go out of control, and become a dangerous attribute. Remember that if taken beyond a point, ambition could not only hurt others but also destroy you.

So keep in mind that the promotions that you achieve, the amount of money that you earn, or the number of friends that you have will be worth it, only if you attained them through hard work and fair means. Keep firm control over your ambition, and draw a line, which you will not cross to attain success.

In life you will find that people or fate can deny or deprive you of much; but one thing that no power on earth can take away from you is your self-respect. It is your most precious possession and you must guard it jealously. If the definition of success is clear in your mind, and your success has not been at the cost of anyone else, your conscience will be clear and your self-respect intact. In this context, let me relate a tale.

I was a Cadet here 42 years ago, and that it is indeed a very, very long time back. Nevertheless, two weeks ago when I got a letter from Khadakvasla that David, my faithful old Cadet Orderly had died, it came as a bit of a shock. My everlasting memory of David was of him fitting cross-legged in the corridor, polishing boots and anklets as he smoked a "bidi". He had a big moustache and I used to tease him that if I ever got checked for bad turnout, I would cut it off. He always replied with a big grin that he would never let that happen.

David invariably knew the training programme, including the changes, and always put out the right uniform. Honest to the core, he was forever busy at work and my boots, belt, anklets and brass-work always shone, and I never got checked for turn out. I don't know how much he earned, but he was always cheerful and contented. He probably contributed to my becoming a Squadron Cadet Captain in the 6th term!

Ungrateful that human beings are, four decades later when I returned as Commandant, David had been deleted from my memory. However, my faithful orderly, now about 80 years old, had not forgotten me. One day he painfully limped to the Commandant's House to greet me and offer his congratulations. Long retired, he would drop in home every month without fail, for a cup of tea, a chat about the old days, and depart with a small tear in his eye and a big smile on his face.

Was David a successful man? I would like to think so. He laboured faithfully for 40 years doing the job he knew best: polishing and brasso'ing, and caring for "his Cadets". He retired as Head Orderly. And his happiest memories were of the Cadets he had served. He was proud that I had come back as Commandant. Before he died he left instructions that I was to be informed. I hope he died a happy and content man.

So at the end of the day gentlemen, if you can stand in front of the minor, look at yourself squarely in the eye and truthfully say that you have done your duty, to the best of your ability, you can certainly claim success. And what is more, if your conscience is clear, your self-respect is intact, you will not merely have been successful in life, but you will also be a happy person too. And that is something to really work for!

VISION OF AN EMPOWERED INDIA

The people of India solemnly resolved on November 26, 1949 in the preamble of the Constitution, to secure for every citizen: Justice, Liberty, and Equality, to uphold the promotion of Fraternity, and ensure the Dignity of the individual. There could not have been a better prescription for empowerment of a people and a nation. Fifty-six years down the line is a good time to undertake a reality check and to perhaps re-order our priorities.

Living in Lutyen's Delhi, I guess one is somewhat on weak ground. But let us first talk about dignity of the individual. Just as nations are empowered by their standing in the world, human beings are empowered by a sense of personal dignity and self-respect. Take these away, and you have an abject amoeba. But that is exactly what happens daily at daybreak, when millions of our brothers, sisters and children awake in slums, in degrading shelters of bamboo and plastic and squat for ablutions beside railway lines, footpaths and public drains.

I have been in cities across the world where if you stand at a vantage point, all you can see is row after row after row of drab, grey housing blocks. Unsightly they may be, but they signify a national resolve to ensure that no citizen will sleep on the footpath. Why do we not empower our citizens by evolving a national consensus that providing basic housing and sanitation, and restoring their dignity as human beings is an issue of top priority?

While abroad, an Indian walks proudly, with his head held high, because he knows that the world knows his worth. Back at home he scurries; eyes averted, nose covered, avoiding piles of uncleared garbage, unruly traffic, and stray cattle. It will not be long before the world too sheds its polite reticence and remarks loudly at the conundrum of India's economic resurgence and intellectual excellence, coexisting with physical

---

Indian Express, October 17, 2005. p. 1. Published in the series entitled "My Vision of an Empowered India"

slovenliness. Every one of us feels diminished at our country's civic neglect, filth, squalor and urban decay. So why do we not empower ourselves by ensuring that the taxpayer's money is spent on enforcing basic civic laws and ensuring cleanliness, and order on the streets. And in getting the ubiquitous garbage off our streets?

Let us talk of Justice and Equality before the law. The common man feels vastly reassured, and empowered when the rule of law is upheld and his environment is safe. But he is continuously watching to see if those who break our laws get their just desserts. He is also looking to see if the enforcers of the law respect it themselves? Why can the guardians of our laws not be selected and trained so that they inspire confidence and respect? Why can we not house them decently and pay them a salary that will give them dignity and self-respect, and protect them from petty temptation. By raising their status in society, we empower ourselves.

Swift dispensation of justice, no matter how exalted the delinquent, is another issue that bothers the citizen. This is today a distant vision and unless our esteemed judiciary adopts some radical measures (some were recently suggested by the President of India), the vision will keep receding, as cases pile up.

It has often been pointed out that as individuals we have quadruple standards of conduct; one for ourselves and one for others; one while in India and another while staying abroad. If our children do not learn civic conduct and a sense of values at home, our educational system should step in to fill the void. Here again we have a conundrum. Our institutions of higher learning are right up there with the best in the world, but our primary and secondary education is in shambles. Let us really, really empower ourselves by investing in our children. By devoting funds, and more importantly, focusing attention on basic education.

What about Fraternity? The greatest gift that the founding fathers of our nation could have bequeathed us, is the precious right of universal franchise, which we have exercised come hell or high water with regularity since Independence. There can be no greater empowerment of a citizen than this. But let us pause here and remind ourselves that elections are not an end in themselves, but merely a means to an end; which is the upliftment of the common man and betterment of our society. In unity lies empowerment of India. Let fraternity and brotherhood amongst our people not becomes a casualty in the rough and tumble of our sacred democratic tradition.

Lest I be called a hypocrite, let me add in conclusion, that the vast sums of money that we spend today on "guns" could certainly be better spent on "butter" for our children or achieving much of what I have said. However, if history has taught us anything, it is that security, not weakness that empowers a nation in the long run. Security also engenders development. But when "Utopia" comes, we shall certainly be ready in biblical tradition to "beat our swords into ploughshares".

FIRST PERSON ACCOUNT

THE USS ENTERPRISE IN THE BAY OF BENGAL
The Real Reason

Lest the title of this story mislead people into thinking that I am attempting to wreck the newborn Indo-US detente, let me start by quoting Henry Kissinger's words from his book, *My White House Years.* He says, "An aircraft carrier task force that we had alerted previously was now ordered to move towards the Bay of Bengal, ostensibly for the evacuation of Americans, but in reality to give emphasis to our warnings against an attack on West Pakistan."

There can be little doubt that the sailing of Task Force 74, headed by the carrier *Enterprise* into India's backyard on December 10, 1971 is something that has rankled Indians ever since. Especially since attacking West Pakistan was not on the agenda. However, with the Henry Hyde Indo-US Cooperation Act of Congress now in place, it is time to iron out these wrinkles. Fortunately, information has come to light recently, which shows that there may have been good and valid reasons for this action by the Nixon administration. And what's more, it also appears that Brigadier General Charles "Chuck" Yeager, USAF, and I may have contributed to this denouement!

For those (very few) readers who may not be familiar with Chuck Yeager, we can say that he is the person who inspired the book *The Right Stuff* by Tom Wolfe, and the famous Hollywood movie of the same name. A few words about this legendary test pilot would not be out of place here.

Yeager enlisted as a Private in the US Army Air Corps in 1941 and entered pilot training to graduate two years later as a flight officer. During World War II, he flew 56 combat missions, in which he shot down 12 German aircraft (including five Me-109s during a single mission). Returning to the USA in 1945, his remarkable flying skills caught the eye of his superiors and he was assigned to the USAF Test Pilot School then at Wright Field.

Vayu Aerospace and Defence Review, 1/2007, pp. 149-52.

On graduation, Yeager was selected as project pilot for one of the most important flight test programmes in history; to fly the rocket powered Bell X-l. On October 14, 1947, after launch from the belly of a B-29, he accelerated to Mach 1.06 at 42,000 feet and became the first pilot to shatter the once dreaded "sound barrier". During his career he flew over 10,000 hours on 330 different types and models of aircraft.

After commanding the USAF Test Pilot School and a fighter wing in Philippines, (flying 127 missions over Vietnam). In 1969, he was promoted to Brigadier General and in January 1971, in his penultimate assignment, he was sent as the US Defence Representative to Pakistan, in Islamabad. And that is where fate decided that our paths should cross.

The cramped and cryptic entry in my flying logbook for December 4, 1971 reads as follows: "Type Flown: Hunter Mark 56-A, No. 463. Mission: 2 aircraft gun strike (Lead) PAF Base Chaklala. Duration: 1 hr 15 mts. Results: 424 rounds HE fired, one C-130 on ground". The preceding entry for the same day says simply: "Delhi-Home Base". And thereby hangs the somewhat unusual tale that I am about to relate.

As a young naval Lieutenant, in the late 1960s, having recently carrier-qualified on the Armstrong-Whitworth Sea Hawk, from the deck of the Indian Navy's sole flat-top *Vikrant*, I was just settling down to polish up my embarked flying skills when I received orders for an exchange posting with the Indian Air Force (IAF). So I packed my bags and with great reluctance, left the sunny beaches of Goa and headed for north India.

Having converted to the British Hawker Hunter ground attack fighter (a second-generation trans-sonic descendant of the Sea Hawk) I was posted to an IAF fighter squadron based close to New Delhi in end-1970. It did not take me long to find my feet on *terra firma*; the air force did everything more or less like the navy, except that they were very serious and professional. But we thought that we performed with greater style and panache, and wore orange 'mae wests' while doing it!

The IAF was good enough to give me a longish spell of leave in mid-1971, during which a friend and I hitch hiked to Europe via Iran, Turkey, Greece and the Balkans. We spent 4th of July amidst boisterous GIs in the Hoffbrauhaus in Munich, and Bastille Day amongst celebrating Parisian crowds on the Champs Elysees. But everywhere we went, the news stands flashed an unfamiliar but ominous new word: "Bangladesh".

I returned to New Delhi to find that my IAF squadron had been moved further north and went looking for it. Our new base was located within 2 minutes flying time (at 420 knots) from the India-Pakistan border and would certainly be an interesting place to be in when the "balloon" went up.

My Squadron "Boss", Wing Commander (later Air Vice Marshal) C.V. Parker, was an unusual combination; a great flyer, and also a martinet. He made it clear that if we ever went to war, he wanted to be sure of two things: (a) that he had prepared us for it in the best possible manner, and (b) that he went in ahead of everyone else. So for the next four months we were put through a most rigorous training programme, focusing on target recognition, low-level navigation, weapon delivery at dawn/dusk and air combat.

The only long-range combat aircraft in the subcontinent in that era were the Canberra light bomber in the IAF inventory, and its US derivative, the Martin B-57 that equipped our rival, the Pakistan Air Force (PAF). The Hunter Mark 56-A that my squadron flew carried four large underwing fuel tanks, which gave it an extraordinary radius of action at low level. Therefore, as far as reach was concerned, apart from the *Canberra*, my outfit had the longest legs in the IAF.

As a single-seat fighter, the Hunter's operating milieu was, however, restricted to daylight hours, whereas night bombing missions were the forte' of the Canberra. It was therefore an unstated sine qua non that wars on the Indian subcontinent would commence only on a full moon night so that the Canberra's and B-57s could be fully exploited.

During summer of 1971, as we watched the sequence of tragic events unfolding in East Pakistan, most of us were convinced about the inevitability of conflict. At the end of the monsoons, to avoid becoming "sitting ducks" for a PAF pre-emptive strike; every full moon phase saw my squadron retiring to a rear base (a few hundred miles away from the international border). Since the December full moon was to be on the 2nd of the month, on November 25 we flew down to Ambala (in Punjab) for some firing exercises and on December 1, 1971 we flew even further south (and east) to Delhi.

Sure enough, on the evening of Friday, December 3, the radio announced that at 5.40 p.m., PAF fighters had carried out a coordinated strike on nine Indian air bases all along the western border. Later that night we clustered around the radio to hear Prime Minister Indira Gandhi tell the nation that we were at war with Pakistan.

Since the die had now been cast, our first task was to get back to base and commence the retaliatory air war ASAP. Having briefed for a 5.30 a.m. take-off, we grabbed a couple of hours of fitful sleep, and tumbled out of bed at 4 on a bitterly cold morning to find the base completely fog bound! Starting up and taxying in blackout conditions would have been bad enough, but the fog made things even more interesting. Some people took a wrong turn on the taxiway and got lost, but my wingman and I were glad to find us lined up for a timely take-off.

It was eerie to be airborne in the pre-dawn dark, flying at 500 feet with a hint of moonlight in the sky and a sheet of fog below. We knew no enemy fighter could be around, but that did not prevent the hair on your neck prickling now and then. We had a safe transit of about 45 minutes to base, save two minor incidents.

As we neared home, one could see in the distance, a very pretty but intense fireworks display. It was the "friendly" tracer, which our local air-defence gunners seemed to be putting up to welcome us back! Some frightfully bad language on RT from the Flight Commander soon put a stop to it.

Shortly after I had touched down on the darkened runway, I saw from the corner of my eye, a green light whizzing rapidly past my port wingtip. It was my wingman who in his excitement, had landed a few knots "hot" and after overtaking my aircraft came back on the centerline ahead of me, luckily missing the runway lights! I heard a muttered apology on the radio, but we had more important things on our minds.

It was still dark as we taxied into our blast pens, and there was just time for a quick wash and bite, while the aircraft were fuelled and armed. Briefing for the first wave of retaliatory strikes on Pakistan was business like and we walked to our aircraft just behind the Boss. I had drawn a two-aircraft mission against PAF base Chaklala, located a few miles South-East of the new capital city of Islamabad. The briefing was to carry out a single pass attack on briefed targets and to look out sharply for enemy Mirage III fighters on patrol, both over target and en-route.

The direct distance to Chaklala was not great, but we were going to do some tactical routing over mountainous terrain and approach from the northwest, so that the radars would not see us till very late. A few minutes into the mission, the butterflies settled down in the stomach as one concentrated on the map, compass, airspeed and stop-watch (which were all the navigational aids one had 35 years ago!). As we approached the target, it became apparent that the fog, which had bedeviled us over north India a few hours ago, was going to spoil our fun again; the sun was still low, and the slant visibility poor, but one could see tall objects and features right below.

Anyway, we pulled up from low level to about 2,000 feet by the stopwatch, and were gratified to see the murky outlines of the cross-runways of Chaklala airfield, but little else. Then a huge tower appeared out of the haze and I thought that the air traffic control would be a worthwhile target for want of anything better. A short burst from my four canon saw the tower collapsing, and as I flew over it, a huge column of water rose to greet me from the debris. Oops! A water tower! I consoled myself with the thought that the PAF would at least go thirsty tonight.

Pulling out of the dive, I desperately scanned the airfield for something more lucrative on the surprisingly bare tarmacs. Suddenly, from the corner of my eye, I spied protruding from a large mango grove, the unmistakable shape of a tall aircraft fin, and a sharply swept-up rear ramp section. A Hercules C-130 under camouflage!

With a bootful of rudder, and the stick hard over, I swung my fighter around and in a shallow dive, hosed the grove with 30 mm shells. A thin wisp of black smoke gave cause for optimism, and I thought that another pass would be necessary. My wingman, keeping a vigilant eye for enemy combat air patrols felt that we were stretching our luck in hostile territory and made his views known.

Pulling out of the second dive, through a gap in the fog I caught a glimpse of a row of small transport aircraft lined up on the secondary runway. The sight was too tempting. Putting all thought of the Hercules transport out of my mind, and ignoring the multiple arcs of tracer fire, I swung around in a tight high-G turn and emptied my guns on whatever was visible of the light aircraft. By now my wingman had lost patience and was yelling on RT. We departed Chaklala at full throttle hugging the deck amidst intense anti-aircraft fire, which seemed to grow by the minute.

Fate was kind, and empty guns notwithstanding; we had an uneventful return passage. We landed back safely at base, feeling elated that we had opened our account and given the enemy a dose of his own medicine. In the de-brief, I concentrated on the hidden Hercules, and other target details, with a passing mention of the light aircraft and skipped the water tower episode altogether.

That evening I heard that Radio Pakistan had complained bitterly about an IAF attack on UN aircraft but decided to ignore it as "enemy propaganda". The Boss, sharp as ever, would however not let go, and for many months after the war, I had my leg pulled mercilessly about the Navy attacking "unarmed neutrals".

The 1971 war had receded into the depths of my memory when last year, I received via e-mail from a young aviation journalist, a copy of an article published in the *Washington Monthly* of October 1985, with the cryptic remark: "You may find this of interest!"

The article, *The Right Stuff in the Wrong Place,* was written by an American diplomat, Edward C. Ingraham. He was political counsellor to the American Ambassador Joseph Farland in Islamabad, when Brigadier General Yeager was head of the Military Assistance Advisory Group (MAAG). Since the article dwelt exclusively on Chuck Yeager, and touched upon events of the 1971 Indo-Pak war, I did find it of interest, and the reason will emerge shortly. However, at this juncture, I must emphasize that the views and opinions that I am going to quote, are entirely Ingraham's and I continue to hold Yeager in great regard for his professional skills.

"In 1971" says Ingraham, "Yeager arrived in Pakistan's shiny new capital of Islamabad to head the MAAG. Yeager's new command was a modest one: about four officers and a dozen enlisted men charged with the equally modest task of seeing that the residual trickle of American military aid was properly distributed to the Pakistanis. All the chief of the advisory group had to do was to teach Pakistanis how to use American military equipment without killing themselves in the process. The job wasn't all that difficult because the Pakistani armed forces were reasonably sophisticated."

He goes on, "One of the perks of Yeager's position was a twin-engined Beechcraft, a small airplane supplied by the Pentagon to help keep track of the occasional pieces of American military equipment that sporadically showed up in the country. Farland, however, had other designs on the plane. An ardent fisherman, he found that the Beechcraft was the ideal vehicle for transporting him to Pakistan's more remote lakes and rivers, with Yeager often piloting him to and fro."

Speaking of the worsening situation in East Pakistan, Ingraham says, "We at the Embassy were increasingly preoccupied with the deepening crisis. Meetings became more frequent and more tense. We were troubled by the complex questions that the conflict raised. No such doubts seemed to cross the mind of Chuck Yeager. I remember one occasion on which Farland asked Yeager for his assessment of how long the Pakistani forces in the East could withstand an all-out attack by India. "We could hold them off for maybe a month" he replied, "but beyond that we wouldn't have a chance without help from outside". It took the rest of us a moment to fathom what he was saying, not realizing at first that "we" was West Pakistan, not the United States."

He continues: "The dictator of Pakistan at the time, the one who ordered the crackdown in the East, was a general named Yahya Khan. Way over his head in events he couldn't begin to understand, Yahya took increasingly to brooding and drinking. In December of 1971, with Indian supplied guerrillas applying more pressure on his beleaguered forces, Yahya decided on a last, hopeless gesture of defiance. He ordered what was left of his armed forces to attack India directly from the West. His air force roared across the border on the afternoon of December 3 to bomb Indian air bases, while his army crashed into India's defences on the Western frontier."

"It was the morning after the initial Pakistani strike that Yeager began to take the war with India personally. On the eve of their attack, the Pakistanis had been prudent enough to evacuate their planes from airfields close to the Indian border and move them back into the hinterlands. But no one thought to warn General Yeager. Thus when an Indian fighter pilot swept low over Islamabad airport in India's first retaliatory strike, he could see only two small planes on the ground. Dodging anti-aircraft fire, he blasted both to smithereens with 20-millimeter cannon fire. One was Yeager's Beechcraft. The other was a plane used by United Nations forces to supply the patrols that monitored the cease-fire in Kashmir."

"I never found out how the UN reacted to the destruction of its plane, but Yeager's response was anything but dispassionate. He raged to his cowering colleagues at a staff meeting. His voice resounding through the embassy, he proclaimed that the Indian pilot not only knew exactly what he was doing but had been specifically instructed by Indira Gandhi to blast Yeager's plane. In his book he later said that it was the Indian way of giving Uncle Sam 'the finger'!"

Ingraham's suggestion that: "To an Indian pilot skimming the ground at 500 mph under anti-aircraft fire, precise identification of targets on an enemy airfield might take lower priority than simply hitting whatever was there and then getting the hell out," was met by withering scorn from Yeager.

"Our response to this Indian atrocity, as I recall," adds Ingraham (tongue firmly in cheek), "was a top priority cable to Washington that described the incident as a deliberate affront to the American nation and recommended immediate countermeasures. I don't think we ever got an answer".

Ingraham says that Yeager's movements and activities during the subsequent conflict remained uncertain, but "A Pakistani businessman, son of a general, told me excitedly that Yeager had moved into the big air force base at Peshawar and was personally directing PAF operations against the Indians. Another swore that he had seen Yeager emerge from a just landed jet fighter at the Peshawar base."

After reading Ingraham's account, and especially after retiring from the Navy, the thought has often crossed my mind that perhaps Yeager had it coming to him from Mrs Gandhi.

And if Indira Gandhi did indeed personally order the destruction of Chuck Yeager's Beechcraft, then Richard Nixon may have been quite justified in personally directing the *Enterprise* task force to sail into the Bay of Bengal as an "immediate countermeasure".

In which case the honours are equally shared, and I owe no apologies to anyone, except perhaps UN Secretary Ban Ki-moon!

INDEX

B

B-29, 216
Baalu, T.R., 111
Baghdad Pact (1954), 3
Baku see *Admiral Gorshkov*
Bali, 47, 81, 152, 167, 176
Balkans, The, 60, 216
Bangladesh, 4, 87, 100, 153, 172, 216
Ban Ki-moon, 221
Battle of Britain, The, 59
Battle of Lepanto, The, 66
Bay of Bengal, 9, 67, 71, 101, 130, 140, 144, 160, 169, 175-176, 215
Beazley, Kim, 165
Beirut, 107
Belfast, 131
Bell X-1, 216
Berlin Wall, 80, 149
Bhagvad Gita, 210
Bhutto, Zulfikar Ali, 4
Bihar, 78
Black Sea, 122-123
Blitzkrieg, 60
Bombay, 75
Boorda, Jeremy, 201
Borobudur, 176
Bose, Subhas Chandra, 141, 159
Bosphorous Strait, 122
Botswana, 58
BrahMos SSM, 94, 110
Brazil, 81, 102
Breguet Alize, 131
British Army, 3
Buddhism, 159
Bush, George W., 166

C

Cabinet Committee on Security (CCS), 14
Cabinet Secretariat (India), 43
Calicut, 48
Calicut University, 45
Cambodia, 47, 68, 81, 152
Canada, 49, 197

Canberra, 197-198
Carnegie Endowment for International Peace, 2, 88
Car Nicobar Island, 139, 184
Carthage, 66
Central Administrative Tribunal (CAT), 208
Central Asia, 6, 26, 50, 99
Central Asian Republics, 72
Central Intelligence Agency (CIA), 18
Centre for Airpower Studies, (CAPS, New Delhi), 2
Centre for Land Warfare Studies (CLAWS, New Delhi), 2
Chaklala (Pakistan), 218-219
Chalukyas, 68, 175
Champa (Siam), 68
Chandragupta Maurya, 175
Cheng Ho, 141
Chetwode, Philip Walhouse, 181
Chief of Defence Staff (India), 15, 23-24, 28-29, 31
Chiefs of Staff Committee (India), 19, 21, 23, 28, 146-147
Chilbolton (UK), 198
China, 4-5, 50, 54-56, 62, 67, 70-71, 75, 81, 87, 99-100, 123, 151, 172, 194
Cholas, 47, 68, 81, 175
Chittagong, 70, 100
Clausewitz, von, 43-44
Clemenceau, George, 1
Clinton, Hillary, 157
Clive, Robert, 48
Coast Guard (India), 54-55, 76, 88, 106, 137, 142-143, 163
Cochin Shipyard, 111-112, 129, 132, 135, 193
Coco Islands, 144
Cold War, 54, 89, 91, 96, 99, 118, 123, 153, 160, 163, 165, 173
Colombo, 42, 90
Columbus, Christopher, 157
Communist Party of the Soviet Union, 80, 149
Confederation of Indian Industry (CII), 74
Confidence Building Measures (CBM), 7, 56
Container Security Initiative (CSI), 55, 118
Crete, 66
Crimea, 128
Cuddalore, 176

M

N